Environmental Values in American Culture

Environmental Values in American Culture

Willett Kempton, James S. Boster, and
Jennifer A. Hartley

The MIT Press
Cambridge, Massachusetts
London, England

Third printing, 1997
First MIT Press paperback edition, 1996

This book was set in Sabon by DEKR Corporation, Woburn, Massachusetts and was printed and bound in the United States of America.

Library of Congress Cataloging-in-Publication Data

Kempton, Willett, 1948–
Environmental values in American culture / Willett Kempton, James S. Boster, and Jennifer A. Hartley.
p. cm.
Includes bibliographical references and index.
ISBN 0-262-11191-8 (HB), 0-262-61123-6 (PB)
1. Environmental conditions—United States. 2. Public interest—United States. 3. Values—United States. I. Boster, James S. II. Hartley, Jennifer A. III. Title.
GE150.K46 1995
363.7'00973—dc20 94-27765
CIP

Contents

Acknowledgments

A work of this scope benefits from a wide circle of colleagues and critics. The authors are fortunate that many have abetted, encouraged, and criticized us. Michael Ross moved the book a huge step forward with a critical review of an early draft manuscript, making major organizational and research literature suggestions. We are also grateful to Steven R. Brechin, Roy D'Andrade, Riley E. Dunlap, Jill Neitzel, Priscilla Weeks, David Wilson, and three anonymous reviewers for their comprehensive reviews of the entire manuscript.

We also appreciate comments on individual chapters by Ann Bostrom, Janet M. Chernela, John M. Darley, Dan Deudney, Glen Ernst, J. Houston Kempton, Lester Milbrath, Apoorva Muralidhara, Vijaykumar Ramakrishna, James Risbey, Robert Socolow, Bron Taylor, and Susan Weller.

Our editor at The MIT Press, Madeline Sunley, provided critical support and advice as we responded to reviewers and finalized the manuscript. The production editor, Sandra Minkkinen, improved clarity and consistency of the manuscript, and Robert E. Schultz created the figures.

In many places, we compare lay people's cultural models of the world with those of scientists. To do so, we draw on the fields of biology, biochemistry, ecology, climatology, atmospheric chemistry, and physics. On our own, we could never cover so many fields at the depths required for these subtle comparisons. We are therefore indebted to several natural scientists for their time and patience—typically in the form of answering endless questions from coauthor Kempton. We acknowledge the help of Al Cavalo, Kerry Cook, David O. Hall, Jerry Mahlman,

Steve Pacala, James Risbey, and Robert Socolow. We are confident that we have at least occasionally garbled their insights and explanations, and remind their colleagues that errors in this book remain our responsibility.

As for data collection, we are grateful to Apoorva Muralidhara, Dan Levi, and Leslie Clark for assistance in conducting the semistructured interviews, and to Anita Iannucci, Jason Masterman, Karston Mueller, and Christina Rojas for their help in administering our surveys. Assistance in contacting informants was provided by Randa Low, William Mautz, Bron Taylor, and Roz Holler (who helped in other ways as well). We are grateful to Bron Taylor for administering our survey to Earth First! members and to Riley Dunlap for providing unpublished U.S. national polling data from his Health of the Planet survey (Dunlap, Gallup, and Gallup 1992; 1993).

This research was made possible by grant BNS-8921860 from the National Science Foundation. We especially acknowledge the encouragement and support of Stuart Plattner and Tom Baerwald, our NSF program officers respectively of Cultural Anthropology and of Human Dimensions of Global Change. Partial support for the first fourteen open-ended interviews was provided by a grant from the Hewlett Foundation to Princeton University's Center for Energy and Environmental Studies. Substantial writing time was supported as part of Kempton's faculty position at the University of Delaware's Center for Energy and Environmental Policy. Major portions of chapter 4, and occasional passages of chapters 5 and 6, draw from reports on our early findings (Kempton 1991a; 1991b). The section on surveys in chapter 1 is also published in Kempton 1993.

Our own division of labor was as follows. Boster and Kempton conceived the project and wrote the NSF proposal that funded most of the data collection and analysis. Kempton developed the semistructured interview protocol; all three coauthors carried out the semistructured interviews, extracted the key ideas for the survey, and wrote up the individual cases (chapter 7 and appendix D). Kempton was responsible for coverage of the science, policy, and environmental sociology literature. Hartley conceived and wrote initial drafts of several sections of chapters 5 and 6. Boster carried out the survey, analyzed it, and wrote

Preface

This book is an anthropological study of how Americans view global warming and other environmental changes. Recent polls show that environmental awareness has increased greatly. For example, approximately half to three-quarters of all Americans now consider themselves to be "environmentalists." This study explores the meanings behind this and other remarkably high poll numbers. Anthropological research techniques, developed for the study of foreign cultures, are used to investigate the beliefs, values, and cultural models that constitute the foundations of public environmentalism. Using several examples of global environmental change such as global warming, ozone depletion and species extinctions, we document how the public transforms scientific information as they interpret it. We also explore American environmental values, and how beliefs and values together influence preferences for or against environmental policies.

Our interviewees include members of the general public, as well as selected groups ranging from radical Earth First! members to laid-off sawmill workers in Oregon. Among the surprising findings are that the public and scientists have completely different understandings of some critical environmental problems and proposed policy solutions, that environmental values have already become intertwined with other American values—from religion to parental responsibility—and that an environmental view of the world is more universal than previous studies have suggested.

Our results provide insights into the nature of environmentalism, insights that many readers will experience as seeing something familiar but understanding it for the first time. Our results also have practical significance, for example, in suggesting ways to greatly improve environmental communication—whether by teachers, advocates, politicians, journalists, or scientists.

chapter 8, most of appendix A, and parts of chapter 1. Kempton conceived and wrote the remaining chapters, and brought the entire draft to the final form. All of us worked on each others' writing, challenged each others' conclusions, and occasionally yelled at each other. Many of the strengths of this volume are the product of these interactions among the authors and with those mentioned above and below.

We close by acknowledging the contributions of our informants; both those who took the time to talk with us in person and those who responded to our written survey. We hope we have lived up to our responsibility as anthropologists to faithfully convey their individual and collective wisdom.

1 Introduction

The natural world is constantly changing. But today's multiple simultaneous changes are unprecedented and, in the view of some scientists, potentially catastrophic. For the first time, the primary driving force of planet-scale change is humanity, with our growing numbers and increasingly disruptive activities. Major global-scale changes include ozone depletion, species extinctions, and global warming. Scientists cannot predict the ultimate effects of these global changes—their scope and pace have no precedents in human history and few precedents in the geological history of the earth.

Along with these environmental changes, the environmental beliefs and values of human cultures are also rapidly evolving. This book, written by three anthropologists, deals with these changing cultural concepts of the environment in the United States. Understanding culture is an essential part of understanding environmental problems because human cultures guide their members both when they accelerate environmental destruction and when they slow it down. For everyone—leaders, citizens, and scientists alike—the cultural framework shapes the issues people see as important and affects the way they act on those issues.

In order to understand environmental perspectives in the United States, we interviewed people from all walks of life. In conducting this research we found that popular environmental sentiment is not an isolated topic but links closely to such diverse areas as religion, parental responsibility, beliefs about weather, and confidence in the government versus industry to solve environmental problems. Reflecting our finding of interconnectedness, this book describes environmental thinking more holistically and more comprehensively than have other studies to date. The result is not just a closer look at environmental thinking but a unique attempt to understand the belief systems and values at the foundation of environmental sentiments in the United States.

Our goal is to analyze the components and causes of popular environmentalism. To do so, we use an anthropological approach. We start with extensive, semistructured interviews with open-ended questions, enabling interviewees to explain their beliefs and values in their own words. These interviews produced surprising findings not detected in previous research. We then constructed a closed-ended survey questionnaire to test how widely our findings apply across diverse groups in American society. The result of these two stages is, we feel, the essential but heretofore undocumented "big picture": the most complete and holistic view yet developed of the beliefs, logic, and values embedded in mainstream American environmental thinking.[1]

Why use anthropological methods in our own culture? In previous research we have compared cultural models regarding energy and environmental issues as seen by differing groups of people. For example, we find vast differences among the cultural models held by laypeople, scientists, and administrators (e.g., Kempton 1987; 1993). Through the process of being socialized into a community of specialists, experts in the science and policy of environmental change can lose touch with lay thinking. By documenting the divergences of lay models from those of specialists, this book can be used to understand why environmental initiatives are supported (or opposed) and to design more effective communication.[2] But we hope that this book's exposition of laypeople will affect environmental specialists in a more profound way. Ordinary people's reactions to current environmental issues sometimes remind us of fundamental values or plain wisdom that can be forgotten in "sophisticated" policy analysis. This book may thus alert specialists when policy goals or analytic assumptions have lost touch with basic values of citizens.

In exploring variation in environmental models, we also hope to make a contribution to anthropology and cognitive science. We find that American perspectives on global environmental change are based on fundamental moral and religious views on the relationship between nature and humanity, other species' rights, humanity's right to change or manage nature, and our society's responsibility to future generations. American environmental views are thus enmeshed in a core set of cultural beliefs and values. Understanding how these core beliefs are structured and how they are distributed within society expands anthropology's understanding of cultural knowledge and cultural values. It also provides a revealing case study for cognitive scientists concerned with the ways in which people assimilate new information by fitting it to preexisting concepts.

Although we cannot speak as authoritatively about possible contributions to other disciplines, we hope that this book also has something to say to the philosopher or ethicist about ordinary people's values and ethical reasoning, to the political scientist about the complexities that underlie voting patterns and public acceptance of policies, and to the environmental sociologist about the structure of American environmentalism. For the student of science and society, we hope to not only show how scientific theories are selected and transformed by laypeople as those theories spread widely through society but also to raise the question of the status of lay science. On a practical level, our extensive examples of discrepancies between scientists' and laypeople's understanding of global environmental problems should be of value to science educators. Finally, since the lay thinking we document seems to explain acceptance or rejection of many environmental policies, we believe that this work will be of value to those who must respond to public opinion about the environment as well as those who seek to influence it.

In the remainder of this introduction, we review measures of the increase in American environmentalism, describe how other scholars explain environmentalism, outline our unique approach, and give an example of a traditional society's cultural models of the environment.

Measures of American Environmentalism

To many readers it may seem obvious that environmental concerns have grown dramatically in recent years. This subjective impression is supported by numerous surveys as well as voting and market data. This section, which briefly reviews the existing evidence of increasing environmental concern, is included so that we do not have to demonstrate it ourselves, thus freeing us to concentrate on our focus: the nature of this environmentalism.

Dunlap and Scarce (1991) review the extensive survey evidence for increased environmentalism among the U.S. public. This section draws examples from that review and elsewhere (surveys not otherwise cited are drawn from the appendix of Dunlap and Scarce's review). Unless otherwise noted, each survey is based on a "national probability sample," meaning that the results are statistically representative of the U.S. population. We use these survey data to make three points: Americans have become significantly more proenvironmental since the sixties, and especially since 1980; their environmentalism goes deeper than just opinion or attitude to core values and fundamental beliefs about the world; and their environmentalism affects market and voting behavior.

Change in environmental thinking can be seen most clearly by the few questions that have been asked consistently over a twenty-year period. For example, the Roper Organization asked whether the respondents thought "environmental protection laws and regulations have gone too far, or not far enough, or have struck the right balance." From 1972 to 1990, those answering "not far enough" climbed from 34 to 54 percent, "right balance" dropped from 32 to 26 percent, and "too far" dropped from 13 to 11 percent. Note that this trend toward increased acceptance of environmental regulation bucked a more general trend of increasing public criticism of other regulations during the same period. Also of interest, "don't know" dropped from 21 to 9 percent as more Americans developed opinions on this topic.

In a second long-term comparison, Cambridge Reports asked respondents to choose between "We must sacrifice economic growth in order to preserve and protect the environment" and the converse. From 1976 through 1990, those choosing to sacrifice economic growth grew from 38 to 64 percent, while those preferring to sacrifice environmental quality dropped from 21 to 15 percent. "Don't know" halved from 41 to 21 percent.

Some more recent polling questions go beyond opinion to personal identity. In 1990, the Gallup Organization asked Americans "Do you consider yourself to be an environmentalist or not?" with the remarkable finding that 73 percent considered themselves to be environmentalists, 24 percent did not. (For brevity, we do not list percentage responses of "no opinion," "not sure," etc. on this and subsequent questions.) Those considering themselves to be environmentalists were about equally split when subsequently asked whether they considered themselves "strong" environmentalists. In a similar question, Cambridge Reports asked respondents in 1990 to mark on a 1 to 7 scale how much they would identify themselves with the label "environmentalist." Fifty-eight percent answered on the "do identify" side of the scale, with 20 percent on the "do not identify" side, and 20 percent in the middle. It bears emphasizing that these questions do not merely ask whether respondents are concerned about environmental pollution (such questions routinely generate agreement above 90 percent). Rather, they ask whether the respondent considers herself or himself "an environmentalist." The majority of Americans now do.

Americans say they want environmental protection even when asked to make difficult trade-offs. A New York Times/CBS 1990 survey found that 56 percent agreed "We must protect the environment, even if it means jobs in the local community are lost"; 36 percent disagreed (Berke 1990). They also say that they will personally pay to help the environment. For example, Yankelovich found in 1990 that 64 percent said "I would be willing to pay as much as 10 percent more a week for grocery items if I could be sure that they would not harm the environment." Only 31 percent disagreed. (In 1971 the figures were 47 and 43 percent, respectively.) Also in 1990, Cambridge Reports asked for a specific dollar figure: "How much more per month would you personally be willing to pay for all the goods and services you use as a consumer, if you knew that as a result . . . business and industry would . . . not harm the environment?" The median response was $36.99 monthly, up from the 1984 figure of $8.10. This increase is not due to inflation—if figured in constant 1990 dollars, consumers' willingness to pay increased from $10.23 to $36.99 monthly.

In addition to opinion polls, data on voting and market decisions reinforce the conclusion that environmentalism is increasing. Voting and purchasing data prove that individuals are willing to do more than answer positively in a survey—they are willing to commit political or financial resources to environmental protection. For example, Hays (1992) summarizes U.S. House of Representatives voting records from 1971 through 1989, which show increasing proenvironmental votes throughout that period. He concludes that "changes took place gradually over the years, were persistent rather than sudden, and point to social, economic and political forces that were incremental rather than episodic" (Hays 1992, 15).

As for market activity, ozone depletion offers what is surely the most dramatic example of a global environmental problem changing consumer purchases. Responding to media accounts during the seventies of the scientific hypothesis that chlorofluorocarbon (CFC) aerosol propellants were depleting the ozone layer, Americans voluntarily switched to nonaerosol propellant alternatives, such as pump sprays and roll-ons, causing a rapid 50 percent drop in market share of products packaged in spray cans (Lyman et al. 1990, 115; Benedick 1991). This consumer response occurred prior to the CFC–ozone depletion link being definitively established, and provided part of the impetus for both the U.S. legal ban on CFCs in spray cans and the subsequent international treaties. In more recent consumer advertising, manufacturers have clearly decided that consumers act on the basis of environmental product claims, as evidenced by the manyfold increase of such claims in recent years.[3] However, such claims by manufacturers highlight environmental features, whereas the product may not be environmentally beneficial overall. By contrast, Germany has established a government-standardized product evaluation and labeling program that is having a substantial environmental impact.[4]

The increase in American environmental consciousness may be observed in language. The same patch of land once referred to as a swamp is now more likely to be called wetlands, and the area once called jungle is now called a rainforest. The process of shifting word choice reflects a change from considering these areas as threatening or disgusting to considering them to be essential components of the ecosystem and precious environmental resources (Ross 1993). Such conceptual shifts, reflected in language, indicate that the changes are not simply a few isolated attitudes but involve a more fundamental cultural change.

Surveys can also demonstrate whether only certain strata of society are environmentally concerned. Studies that have looked only at activists in environmental groups often find their membership to be better educated, wealthier, and less ethnically diverse than the general population. Such characteristics, however, are typical of activists in many political causes. We need to ask if these sociodemographic variables explain environmental sentiment among the broader general public.

Several studies of public environmental sentiment show that it is not strongly related to social elites (Mohai 1985, 1990; Morrison and Dunlap 1986). A comprehensive summary of a large body of literature by Mohai and Twight (1987) concludes "Recent improvements in data and data analysis have shown that concern for the environment is and has been much more broad-based than previously believed, cutting across most, if not all, indicators of socioeconomic position." Their finding thus contradicts a misconception still heard in university seminars—that public environmental sentiment is limited to people who are educated, wealthy, white, associated with universities, or politically liberal. This may be true of activists, and it may have been true of the general public when modern American environmentalism began as a minority sentiment (Sills 1975). However, since the eighties, the only social variable strongly and consistently correlated with public environmentalism is age, with youth's greater concern with environmental issues indicating that the United States is continuing a historical shift toward a more environmentalist population (Mohai and Twight 1987). Even among activists, the association with social class breaks down for local environmental movements, many of whose members are poor or from minority groups (Bullard 1990; Bryant and Mohai 1992).

Taken together, the evidence outlined in this chapter leaves little doubt that a substantial change is taking place in the way Americans conceive of the environment. Environmental activists may not be satisfied with the pace of the changes or their practical effects. Nevertheless, the fact that the indicators we reviewed include voting records and market purchases means that the proenvironment sentiments carry some force—they are not just empty platitudes. Further, they are widespread across strata of society, and their increasing strength among the young portends a continuing shift in this direction.

What Is Environmentalism?

What is the nature of this increased environmentalism? Why has direct public action had strong effects on some environmental problems, such as CFCs and ozone depletion, but little effect to date on others, such as global warming? If it is a shift in culture and values, how fundamental is it really? To put these questions another way, when 58 to 73 percent of the population say "I am an environmentalist," what could they be referring to?

Other scholars have inquired into the nature of environmentalism; we review some of those works here, dividing those who see a primary basis for U.S. environmentalism from those who see multiple causes and multiple elements.

Many of the explanations for environmentalism have hypothesized a single cause or motivation. For example, Douglas and Wildavsky (1982) attribute American environmentalism to the activities of environmental activist groups. Such groups have selected which dangers to worry about—environmental harm rather than unemployment or crime, for example—not based on dispassionate risk analysis but in order to justify and perpetuate their own organizations. Thus, for Douglas and Wildavsky, environmental groups are a cause, not a result, of increased public environmental concerns.

Tucker (1982) also sees environmental activist groups as a primary driving force. He explains environmentalism as driven by an influential elite of activists demanding sacrifices in material wealth to achieve marginal societal gains. Tucker says that one of the environmentalists' key motivations is to preserve wilderness areas for their own exclusive use. Hays (1987) sees a similarly utilitarian motivation behind public environmentalism—he identifies immediate desires for health, scenic natural environments, and quality of life as key bases for support of environmental protection.

Each of the above authors focuses on a primary component, such as fear, or a primary cause, such as persuasive activists. This book definitively establishes that single-component views of public environmentalism are grossly oversimplified. We feel that our data, although indirect, render single-cause explanations implausible as well. Thus we see more value in addressing prior research that takes a multifaceted view of environmentalism.

Dunlap and Van Liere (1978) argue that in the seventies, Americans were in the process of developing a "new environmental paradigm," including such beliefs as the fragility of nature, natural limits to growth, need for environmental protection, and desirability of a steady-state economy. They use "paradigm" in the sense of Kuhn (1962) to denote a comprehensive way of seeing the world, incompatible with previous paradigms. Henderson (1976) and Harman (1977) had speculated that environmentalism represented a new paradigm, and Dunlap and Van Liere constructed a survey instrument to measure the beliefs it comprises.

Milbrath (1984; 1989) also sees a new environmental paradigm, advanced by a "vanguard" of environmental activists but with a majority of the public as "environmental sympathizers." He adds environmental values to the Dunlap and Van Liere survey instrument. Olsen, Lodwick, and Dunlap (1992) more broadly hypothesize that the new environmental paradigm is just one component of a "post-industrial worldview." This more general set includes, for example, the beliefs that technology can do as much harm as good, that government's function is to promote well-being, and that smaller organizations are more simply structured. Similarly, Inglehart (1977) sees environmentalism as a prime example of the more general development of "post-materialist values." These values have emerged in industrialized countries only after a generation has grown up with their material wants satisfied and can then turn to nonmaterial concerns such as the environment.[5]

The authors using terms such as *paradigm* and *world view* claim that the environmental vision is a fundamentally different way of viewing the world. For example, Cotgrove characterizes meetings between holders of the environmental and traditional paradigms as follows: "Protagonists face each other in a spirit of exasperation, talking past each other with mutual incomprehension. It is a dialogue of the blind talking to the deaf. Nor can the debate be settled by appeals to the facts. We need to grasp the implicit cultural meanings which underlie the dialogue." (Cotgrove 1982, 33). This is our perspective as well.

Like all the above authors, our goal is to analyze the components and causes of popular environmentalism. We differ in that we break the paradigm into chunks and examine each separately, using methods quite different from those of other scholars. As we describe in the next section, most of this book focuses on cultural models, returning to paradigms near the end. Cultural models cover a limited, specifiable domain and have a well-defined set of methods for eliciting them.

Our methods allow us to go further than prior work in describing the logic and fine structure of environmentalism. The resulting findings support prior studies that consider environmentalism to comprise an interconnected set of values and beliefs. However, we are critical of some previous methods, such as questionnaires using lists of interview questions formulated by the investigator, methods that assume that the investigator knows in advance which values and beliefs comprise environmentalism. Such collections of questions have been inspired by the writings of environmental advocates and analyses of advocates' worldviews. Those previous studies seem to embody a contradiction between inferred causes and methods. Researchers taking a paradigm view reject activists as the root cause of public environmental sympathies. Yet we see in their selection of questions precisely those questions that would be chosen by environmental activists. As we describe in chapter 2, we use methods that assume that we are ignorant of the components of environmentalism, so we must rely on our informants to enlighten us.

Our Approach: Models and Variation, Beliefs and Values

Our approach to American environmental thinking exploits two central concepts from cognitive anthropology: First, people organize their culture's beliefs and values with what we call *mental models* or *cultural models*. Second, agreement or disagreement about these cultural models often has a clear social *pattern of variation,* as can be shown by analyses of which beliefs and values are shared across which groups of society.

The term *mental model* refers to a simplified representation of the world that allows one to interpret observations, generate novel inferences, and solve problems. This concept derives from prior studies of learning, perception, and problem solving (Johnson-Laird 1980; Clement 1982; Gentner and Stevens 1983). In the process of learning, people do not just add new information to a loose accumulation of facts in their heads. Rather, like scientists theorizing, they construct mental models that make sense of most of what they see. Then people can use these models to solve problems or make inferences, based on seemingly incomplete information. On the other hand, models can lead people astray, as when students and teachers have different models, a common problem in the classroom that neither group may recognize (McCloskey 1983).

The contrast between a loose accumulation of facts and a mental model can be illustrated by the popular assimilation of Rachel Carson's message in *Silent Spring* (1962). We do not believe that readers remember only a list of facts—for example, that because of DDT, eagles have trouble laying eggs and the beach is littered with dead pelicans. They go further and construct general models—that different forms of life are connected or, perhaps more specifically, that toxins concentrate as they move up the food chain. A related example is our own finding that Americans now have extended from observations like those of Carson to more general models of the interconnectedness of all species and the possibility of what some call "chain reactions" in nature. We demonstrate that such models give an underlying structure to environmental beliefs and a critical underpinning to environmental values.

What is the difference between a mental model and a cultural model? Psychologists refer to "mental models" as held by individuals (Gentner and Stevens 1983). Anthropologists more often focus on models that are shared within a culture or social group, and thus refer to them as "cultural models" (Holland and Quinn 1987). People in the same culture often construct the same models, even though many fundamental mental models are never discussed explicitly. In this book, our direct data are the *mental models* of the individuals we interview, which, when we find them widely shared, we argue are American *cultural models*.[6]

The second key aspect of our approach is determining the extent to which cultural models are shared. To establish this, our interviews spanned a broad range of interest groups ranging from members of the radical environmentalist organization Earth First!, encompassing the middle ground of ordinary citizens, to managers and employees of polluting industries and laid-off sawmill workers in the Pacific Northwest. From the public debate and media coverage of environmental controversies, one would expect great divergence across these groups. Our study's surprising (and encouraging) news is that despite some differences when comparing opposite extremes, there is a remarkably strong consensus across this wide spectrum on a core set of environmental values.

Why do we study environmentalism through these target groups? A representative national sample would be essential to measure the prevalence of environmental opinion in the country, but that is not what we want to do. Our eclectic sample—giving excess weight to both extremes—is more useful for probing the structure, limits, and invariant bases of U.S. environmentalism. When our methods yield unexpected and pivotal key concepts, we hope our work will intrigue survey researchers to add questions to their polls, thus providing a more precise measure of the national prevalence of these concepts.

Crosscutting our use of models and variation is our inclusion of both beliefs and values as objects of study. We use the word *beliefs* to refer to what people think the world is like and *values* to refer to their guiding principles of what is moral, desirable, or just. Either beliefs or values may be incorporated into a cultural model or may stand alone as simple isolates. A knowledge of existing American beliefs is crucial even for readers concerned with motivation or action, because one of our essential findings is that these beliefs partially determine which environmental issues people attend to and act upon, and what environmental policies they support.

Similarly, values are crucial, especially in the area of global environmental change, because the worst consequences of global change will be experienced in the future. If all people intently pursued their individual economic self-interest, based on their own past experience, nothing would be done to improve the situation. Cultural values are a necessary basis for environmental action, even if they may not be sufficient by themselves.

What would motivate people to take action when it cannot benefit them directly? Our interviews reveal that Americans are motivated by a diversity of values, ranging from religion to human utility and even to a conviction that nature itself has rights. For example, one strong value we found was a feeling of responsibility for one's children and subsequent descendants. For those who have children, the anchoring of environmental ethics in responsibility to descendants gives environmental values a concrete and emotional grounding stronger than that of abstract principles.

That environmental values are already intertwined with core American values, such as religion and parental responsibility, is a major finding from our research. As this book fills out the details of these interconnections, it will help explain why people who may otherwise be preoccupied with short-term self-interest are now concerned about long-term environmental change.

Non-Western Models of the Environment

In reading drafts of this book, readers of varying backgrounds have reacted in very different ways to our description of American models of the environment. Scientists often react with amusement or frustration to what they consider misconceptions about their specialty. (Of course, outside their own specialties, scientists share the same cultural models as everyone else.) Nonscientists react with "Of course that's what Americans believe. That's the way nature actually works—everyone knows that." Why these reactions? In interpreting our findings in terms of their specialized research literatures, the scientists see discrepancies. On the other hand, the nonscientists are reading about their own culture's fundamental assumptions and see our results as obvious or trivial. In their reactions, both the scientists and nonscientists are missing our point: everyone believes their own models of the world are correct because this belief is continually reinforced by interacting with people who share the same cultural models and use them in the same ways—whether in the lab or at home. To provide some perspective, we describe one very different set of cultural models of nature.

Our example draws from the Tukano Indians in the Northwest Amazon, as described by Janet Chernela (1982; 1985; 1987). The Tukano live along the Uaupés River, a river so poor in nutrients that its fish population would not survive if not for a twice-yearly flooding, during which they swim up into the adjacent forest and gorge themselves on forest food sources. The fish also spawn during this period. Persistence of this ecosystem has been essential to the Tukano, who depend on the fish for protein. Tukano kinship groups (clans) are governed by a guardian spirit or "first ancestor," each with a proper territory or "sitting place" that must be respected. The Tukano apply their kinship model to the fish, also divided into clans, each with a guardian spirit who, like their human counterparts, will be vengeful of transgressions against its clan's sitting place. An offending fisherman's children are expected to be born deformed, whereas fishing in the proper places is perfectly acceptable to man, fish, and spirits.

For one Tukano area that Chernela investigated in detail, guardian spirits were said to protect twenty-six sitting places that would otherwise be potential fishing areas. She found that these areas were not arbitrary but coincided with the forest edge that floods during high water season, as well as islands that served as spawning grounds. Similarly, because the forest edge belongs to the fish even when not flooded, Tukano never farm or clear there. In short, the proscribed locations are precisely those that are critical to leave alone for the fishes' continued viability.

We wish to make two points with this example. First, a cultural model of fish clans, rightful sitting places, and vengeful guardian spirits is a very different way of modeling the environment from that of Americans. Second, this cultural model is a highly effective social control for maintaining a nutrient-poor ecosystem. Its effectiveness may in part be attributed to its weaving beliefs about fish with a moral prescription of the proper relationship of man and nature and a corresponding threat of punishment. Western ecologists' scientific models may be more publishable in biology journals, but those scientific models would be unlikely to guide and restrict these fishermen as effectively.

Unfortunately, as is the case for most traditional cultures, the system we describe is being rapidly destroyed. Tukano land is under pressure from developers who want to use the river edge for agriculture. Even the transmission of their cultural models is threatened. In towns with schools, well-meaning Western-trained school teachers tell Tukano children that their parents' beliefs about the fish and forest are ignorant superstitions (Chernela 1987).

Examples like the Tukano are common in the anthropological literature (e.g., Rappaport 1968; Peña 1992, 81) but are not universal. Some environmentally related beliefs seem to have no known function, and some traditional peoples have been environmentally destructive.[7] Nor do we mean to say that American environmental beliefs have the same origins as those of the Tukano. As we show in this book, in the United States laypeople create models by mixing parts of the scientists' models with parts that laypeople construct themselves.

We hope this non-Western example can also serve as a reminder later in the book. We sometimes compare cultural models of the American layperson with models drawn from scientific literature. These comparisons show that American cultural models are not the only way to conceptualize the world. To say that the scientists' model is different from the non-scientists' is not to say that either is better. As with the Tukano, a cultural model may be functionally superior to a scientific model in the specific contexts in which members of the culture apply it.

We find it somewhat ironic to be using the term *layperson* to refer to the general public. The word *lay* originally contrasted with the clergy, a group who, like scientists, represent themselves as authorities about the truth. Laypeoples' lack of given authority does not mean that they passively receive cultural models. Rather, they actively construct models, possibly building from snippets of popularly reported science or from local wisdom of friends and family. And their ultimate test of a model is practical—whether it works to explain everyday life. Such characteristics of a model's origin, transmission, and evaluation in use, not any characteristics of the models themselves, distinguish a layperson's cultural model from a scientist's model.

Structure of the Book

The following chapters in this book present our findings on the environmental beliefs and values of Americans. Chapter 2 provides background on global environmental change and our research methods. The subsequent chapters report the findings from our interviews.

The topical chapters begin with chapter 3, which deals with cultural models of nature—specifically, interrelationships among living species and humanity's dependence on nature. Chapter 4 explores cultural models about weather and the atmosphere. We show how people use those models to interpret environmental news, to make their own weather observations, and to draw conclusions about the hole in the ozone layer and global warming. Chapter 5 considers Americans' environmental values and the reasoning supporting those values. Chapter 6 examines lay reasoning on policies and institutions to deal with global warming. We show how policy preferences, far from being isolated opinions, can be partially explained by the cultural models and values explicated in earlier chapters.

Our integrative chapters begin with chapter 7, presenting case studies of four individuals selected from the semistructured interviews. The purpose of picking these cases—influential specialists addressing environmental policy—is that in arguing their positions, they reveal more explicitly how individuals select information and construct a coherent view of the world. Chapter 8 explores the distribution of environmental beliefs and values through diverse subgroups, ranging from Earth First! to sawmill workers, in order to discover what is generally shared by Americans and what is idiosyncratic to extreme groups. We outline our conclusions in chapter 9.

Four appendixes provide more detail for the interested reader. In appendix A, data collection and analysis are described in greater detail than in the summary background in chapter 2. Appendix B provides demographic background on the informants from the semistructured interviews and the survey. For the semistructured interviews, it matches pseudonyms with basic demographic data, which some readers may find helpful to refer to in interpreting the informant quotations throughout the book. Appendix C lists the questions used in both the semistructured interview and the survey. Appendix D presents case studies, like those in chapter 7, but of citizens rather than influential specialists.

2

Background

This chapter describes our anthropological methods, then summarizes the science and policy issues behind three major problems in global environmental change: species loss, ozone depletion, and global warming. In each case, we focus on those aspects needed for the reader to understand our research findings on laypeople's views.

We originally embarked on a study of global warming, but we found that interviewees often discussed other environmental problems at the same time, and they drew from the same set of beliefs and values to interpret them. There was an especially large overlap with the three global problems we discuss in this chapter—species loss, ozone depletion, and global warming—which have many common characteristics. For example, these issues exhibit a complexity that makes them challenging to understand, and because their effects are spread out in time and space, they are ethically more difficult as well. These shared characteristics and the similarities seen by our informants led us to add species loss and ozone depletion to several chapters of the book, although the primary focus is on global warming.

Our Anthropological Methods

Our study is unusual within environmental research in that it uses methods developed by cultural anthropologists for understanding foreign cultures. Even among anthropological studies, it is atypical in combining personal interviews with more formal and quantitative methods for analyzing cognitive variation within a culture. This section provides background on both methods so the reader can understand our resulting data and interpretations.

Methods from Cognitive Anthropology

Cultural anthropologists study other cultures. Cognitive anthropology is a branch of cultural anthropology that examines how communities of people come to share cultural understandings of the world: how people reason, make decisions, construct culturally shared models of the world, and act on the basis of those models, and how cultural knowledge comes to have a patterned distribution through society. In short, cognitive anthropology tries to understand how culture emerges in social groups.

The anthropologist learns about a foreign culture using *ethnography*—a suite of methods that include observation, participant observation, casual conversations, and formal interviewing. As cognitive anthropologists studying our own culture, we use two methods developed in the study of other cultures. One is qualitative: *semistructured interviews* and textual analysis of transcripts from those interviews. The other is quantitative: a fixed-form *survey* and analysis of variation among people and groups in their written responses to the survey. (Surveys are used in many disciplines, but our variation analysis of the survey responses derives from anthropology.) These two methods offer different strengths, generally practiced by contending factions within anthropology. They are rarely used together in a single anthropological study.[1] The textual analysis of the semistructured interviews yields rich insights into people's environmental beliefs and values, which are often very different from those of environmental scientists. The cultural variation analysis of the survey delineates the distribution of these beliefs and values among diverse groups within U.S. society. The synthesis of these two methods synergistically yields more insight into American environmental thinking than would either method used independently.

Our contribution to an understanding of environmental concern is not in showing that it exists, a task already accomplished by the national polls, but in documenting the reasoning behind that concern. National surveys are like satellite photos, giving a broad overview of public opinion. Anthropological research corresponds to exploration on the ground, charting details of the features glimpsed by the national surveys and looking for causal explanations. In principle, nothing would prevent us from using our detailed methods on a large national probability sample, but, like trying to explore a continent on foot, it would be impractical to do so.

Our sampling strategy differs from survey research just as our goals do. Whereas a national survey would draw a large probability sample to establish exact percentages of public opinion, our sample includes laypeople as well as members of carefully selected groups who have divergent work or personal interests in environmental issues. These specialist groups include congressional staff, environmental activists, coal workers, industry lobbyists and so on. We chose these groups because they bracket the range of variation in environmental views across the society. Their diverse interests also illustrate how individuals construct environmental beliefs and values consistent with those interests. The chosen specialist groups were different for the qualitative and quantitative parts of the study.

Semistructured Interviews

We began with *semistructured interviews,* in which the bulk of the interview is guided by a protocol of written questions that were asked aloud. The questions encourage paragraph-length rather than word- or sentence-length answers, and the informants are given leeway to elaborate or even bring up new topics they consider relevant. We also create probe questions on the spot to pursue topics raised by informants and paraphrase what the informant says in order to verify or correct our interpretations. In semistructured interviews, probe questions are essential to understand what answers really mean and to explore unfamiliar or unexpected concepts. Interviewees are called *informants* rather than respondents due to their active role and because in most anthropological fieldwork situations they know more about the subject under discussion than the interviewer. The semistructured interviews were conducted by the authors of the book, typically working in pairs or with a research assistant. (See Bernard 1994, 209 for a recent textbook discussion of these interview methods.)

Our interview protocol began with several questions about the weather, then general questions about environmental protection, and specific questions on global warming. We referred to global warming as *the greenhouse effect* in the interviews, since that term has been used more in the media. We then gave a briefing—a short presentation describing global warming—and asked for reactions to it and to several policies proposed to deal with climate change. Except for reactions to policies, the bulk of our data was taken before the briefing; when the briefing could affect the interpretation, we have been careful to identify informant quotations that were taken after it. The semistructured interviews were taped and transcribed in full.

Direct quotations from the semistructured interviews are interspersed throughout the book, letting informants speak in their own words. We follow the guideline that a quotation is included only if similar points were made by several people. An exception was sometimes made for what might be considered antienvironmental statements, which we often included even if expressed by one person. This is because such statements are less common, yet we wanted to discuss both sides of issues. In quotations, our interchange with the informant is conveyed by using italics for our questions and regular type for the informant's response.

Forty-six informants participated in the semistructured interviews. They included twenty lay informants as well as approximately five each from four specialist groups: grassroots environmentalists, coal industry workers, congressional staff working on environmental legislation, and automotive engineers. The nonspecialists were from New Jersey and Maine and were picked to represent a diversity of social backgrounds; two acquaintances were selected because they had expressed antienvironmental opinions, again to insure that such views were represented. Of the specialist groups, the coal workers were chosen with the expectation that they would oppose clean air legislation, which they considered to be threatening coal jobs. We chose congressional staff representing factions both favoring and opposing strong measures to avert global warming. Finally, automotive engineers were included because they play an important role in determining how society uses energy. This diversity of groups insured that our semistructured interviews would elicit a wide range of the beliefs, values, and cultural models which our society applies to the environment.

Overall, the semistructured interview sample had an average age of forty-three, median income of $35,000, and had completed an average of fifteen years of school. They were 67 percent male and 46 percent Republican voters.

Fixed-form Survey

Whereas semistructured interviewing allows the informant to express freely his or her understanding of an issue, the complementary strength of fixed-form surveys allows us to discover how widely distributed those individual understandings are. When only one informant expresses a particular idea, the researcher using semistructured interviews cannot tell whether that individual alone has the idea or whether others share it but do not volunteer it.

To construct the fixed-form survey, we extracted informants' quotations from the transcripts of the semistructured interviews, with editing to make them understandable outside the context of the interviews. A few questions were also added to fill in areas that seemed important to us. We presented these statements to a second set of respondent groups, and each respondent indicated their degree of agreement or disagreement with the propositions.

We administered the survey to five groups. They included two groups of environmentalists: members of a local chapter of the Sierra Club in Southern California, and participants in two Earth First! meetings, in Vermont and Wisconsin. The Sierra Club is a large national environmental organization, with a half-million members and a $35 million annual budget. It lobbies the government on environmental issues and recruits members by direct mail. Earth First! has a loose organization that carries out direct action for protection of wilderness, sometimes including sabotage of logging equipment and actions in which demonstrators place themselves in physical danger. It is estimated to have roughly fifteen thousand adherents (there is no formal membership status) and a $200,000 annual budget (Mitchell, Mertig, and Dunlap 1991; Mitchell 1989). Earth First! was formed partly out of frustration with the slow progress of reformist mainstream environmental groups. Those who have studied Earth First! typically call it a "radical environmentalist" group, and several have noted a religious dimension to its adherents' environmental beliefs (Devall 1991; Taylor 1991, 1993).

To elicit opposing viewpoints, we also surveyed members of groups whose livelihoods had been hurt by environmental legislation: managers of dry cleaning shops in the Los Angeles Basin, whose businesses have been affected by local clean air legislation controlling the release of organic solvents into the air, and laid-off sawmill workers in Oregon. The sawmill workers were expected to be more antienvironmental because their job loss may in part be a consequence of environmental restrictions on timbering and because environmentalism has been shown to be lower for people who are in resource extractive industries such as the timber industry (Cotgrove 1982). In addition, we surveyed members of the general public in California. In each of those five groups, twenty-five to thirty individuals were interviewed. As a cross-check on our two methods, we also surveyed many of the informants who had participated in the original semistructured interviews. This sixth group was not included in most of our analysis.

Two potential improvements to this sample were precluded for practical reasons. First, the ends of our spectrum of groups might be more comparable if the antienvironmental group were more ideologically defined than the sawmill workers, but such groups are exceptionally rare.[2] A second possible improvement would be to use larger samples and probability sampling for each group. This would have increased our confidence in comparing groups and in generalizing from our samples to each population. However, probability sampling was impractical for some subgroups (for example, Earth First!). Although it could have been done for some subgroups, such as the general public, we judged the gains to not warrant the delay.

To briefly characterize the survey sample, the five groups totaled 142 people, with an average age of thirty-nine. They were 67 percent male (disproportionate by gender partly because sawmill workers and dry cleaning managers were predominately male). They averaged fourteen years of school, and 32 percent belonged to an organized religion.

How to Interpret Our Interview and Survey Data

This section presents illustrative data from the semistructured interviews and from the fixed-form survey. It is included to help in interpreting data presented throughout the book. In addition, two quotations from the semistructured interviews are used as an example of how the findings from the semistructured interviews were used to construct the survey questions.

These quotations illustrate an important and widespread concept: "chain reactions" in nature. We did not ask about this concept, but it was volunteered in informant descriptions such as the following:[3]

> Life has a chain, and when they destroy insects, they destroy bird life, and destroying bird life destroys other life. That's a chain reaction, for all of that . . . from the smallest microscopic life [on up].—Walt (retired machinist)

> [With global warming] everything will just get all out of whack. I think once you start mucking with something that big, it will chain react through everything that we're used to in unpredictable ways.—Pervis (coal mine wireman)

Quotations like these from the semistructured interviews give the informant's pseudonym and occupation. For those informants we selected because of their environmental activism, we add environmentalist after their occupation. This identifying information serves to contextualize the quotation—in the case of Walt and Pervis above, that those volunteering the concept of chain reaction are not restricted to just environmentalists or those in professional occupations. For some readers, the name and occupation may also serve as a memory jog for connecting to other quotations by the same informant elsewhere in the book.

Working from transcript quotations such as the above, we reword the ideas into a single line of argument for the fixed-form survey.

105 Nature has complex interdependencies. Any human meddling will cause a chain reaction with unanticipated effects.

Earth First!	Sierra Club	Public	Dry cleaners	Sawmill workers
97	89	77	76	63

The number in front of the question represents its order in the survey. The survey results are always disaggregated across the five survey groups, as above. The numbers represent the percentage in each group agreeing with the statement, counting all who agree to any degree, whether strongly or somewhat.

Given our small and nonrandom samples, how should these percentages be interpreted? Often, as in this case, the results are clear. Even among the sawmill workers, whose own jobs could be interpreted as human meddling with nature, over 60 percent agree with this proposition. Three-quarters of the public agrees, as do at least 90 percent of environmentalists. The statement is clearly a mainstream position, even if the percentages are imprecise. (Some survey statements differ more among groups, as we discuss in our chapter on patterns of agreement.)

When possible, we also validate our results with corresponding questions from national probability samples. In general, few surveys have asked in depth about how nature works because most survey researchers consider what they are measuring to be opinion, not cultural models. No prior survey has asked about chain reactions as we do. However, a recent survey did ask about a related concept, whether "the balance of nature is very delicate and easily upset by human activities." (Dunlap, Gallup, and Gallup 1993). Among their national probability sample, 86 percent agreed and 11 percent did not. This is an example of how we use existing surveys of national probability samples, when any exist with comparable questions, in order to verify our own in-depth results from our more limited samples.

Our statements on the survey often try to capture multipart pieces from the semistructured interviews, such as a cultural model or a chain of thought. Because of this, some statements on our survey are worded in ways that are not standard procedure in surveys—they may seem to be testing several things at once or may not seem neutral. These types of questions are intended and are taken into consideration in our interpretation of the results. For the interested reader, appendixes A through C provide more information on our survey questions, our interviewing and analysis methods, the samples, the pseudonyms used in the quotations, the protocol for the semistructured interviews, and the full survey with answer frequencies.

Global Environmental Changes

Environmental changes are occurring to many natural systems, on local and global scales (Turner et al. 1990). Although we initially focused on global warming, ozone depletion became an important topic because we found that our interviewees had already developed a cultural model of it, which they used to understand global warming. Likewise, species extinction is included as a related issue because it is a consequence of global warming and because informants brought it up in discussing the value of nature.

These three global changes also have important characteristics in common. Unlike local environmental problems, they typically do not damage the immediate local area. For example, after local habitat destruction, plant and animal populations typically recolonize from adjacent populations if land is later allowed to regrow. Ozone-depleting chlorofluorocarbons (CFCs) and greenhouse gases such as carbon dioxide (CO_2) are not toxic to humans or other species. CO_2, part of the photosynthesis and respiration cycles taking place between plants and animals, occurs naturally and is essential to the biosphere. Even though it is causing climate change globally, locally it is not even pollution. These are problems less of destructive acts or substances than of the scale of human perturbations to the planet.

Trying to deal with the problems of global warming, ozone depletion, and species extinction will present a complex set of policy issues. This section first defines the policy aspects common to all three of these problems. It then describes each of the three problems individually.

Environmental Policy Issues

Policies to deal with any environmental problem, local or global, can aim toward either prevention or adaptation. *Prevention* is the method most often thought of for solving environmental problems. If a human activity is causing an environmental problem, a preventive approach would either modify or stop the activity so that the problem would not occur. By contrast, *adaptation* is the decision to find ways to live with a changed environment without trying to prevent the change. For example, on rivers in the states of Oregon and Washington, large dams block young salmon swimming from their spawning grounds to the ocean, despite the aid of fish ladders. A prevention policy would be to not build more dams or to deactivate existing ones. One adaptation policy would keep the dams running, but financially compensate the fishermen who depend on salmon. Another adaptation policy, currently practiced, is to carry some salmon around the dams in barges to try to reverse their precipitous population decline.

Solutions to environmental problems, especially global problems, must overcome the disconnection between cause and effect. The people causing the problem are often remote from the people or other creatures suffering effects. In the case of urban air pollution, for example, the people who drive cars or operate polluting equipment may not be the same people who suffer from asthma or other pollution-related health effects. The economic costs of health care are spread out and not connected to the costs of driving cars or other polluting processes. Even if a particular car driver suffers from pollution-exacerbated asthma, reducing his own car's emissions yields no measurable direct benefit—his asthma persists due to the continued pollution from all the other cars. This disconnection between the causes and effects means that prevention of environmental problems must involve either altruistic action on the part of many individual polluters or collective sanctions, such as the government's passing antipollution laws or pollution taxes. The disconnection problem is worse for global environmental change because the damage to people and the environment can be even more distant—occurring in other countries or in the future beyond our own lifetimes.

The trends underlying global environmental problems are increasing population, high levels of consumption and of waste in industrialized countries, and the spread of Western-style economic development. Currently, population and economic development are expanding most rapidly in developing countries. For example, the third world will soon displace industrialized countries as the major source of greenhouse gases, primarily due to their increased consumption of fossil fuels (Flavin 1992). This shows clearly that global environmental problems, such as global warming, ozone depletion, and species extinction, cannot be solved only by working in the developed countries. Solutions must include facilitating environmentally benign development of the poorer countries (World Commission on Environment and Development 1987) while improving equity across countries (Byrne, Hadjilambrinos, and Wagle 1994).

Habitat Destruction and Species Loss

One can readily visualize a hunter killing the last member of a species. It is more difficult to see how species are lost when construction of a shopping center or a tract of homes shrinks some habitat below a critical size. Yet today, far more species are lost through destruction of the habitats upon which they depend than by killing them directly. Humans destroy habitats by clearing land for farming, grazing, and settlements; by draining and filling wetlands; and by poor land management, leading to desertification. Humans also destroy species by pollution and by introducing new species that overtake native species.

An expected future human cause of extinctions is global warming (Lester and Myers 1989). Climate usually changes gradually, and plants and animals shift north or south to maintain the climate conditions they require. But human-caused global warming will occur rapidly (unlike most natural changes), and human settlements, agriculture, and transportation systems have created many barriers to migration. Both the speed of climate change and the barriers will increase extinctions from the climate changes we induce.

Extinctions occur naturally, at an estimated background rate—without human intervention—of less than one per year. With human-caused changes, current rates of extinction are estimated to be somewhere between four thousand and twenty-seven thousand per year (Wilson 1989; World Resources Institute 1992, 128; Peters and Lovejoy 1990). In the earth's history, such massive extinctions have marked the transition from one geological age to another (Wilson 1992; Simberloff 1986).

Far more species perish per unit area lost of tropical rainforest than of temperate zone ecosystems. An estimated 50 to 90 percent of the world's species are in tropical forests (World Resources Institute 1992, 128), many existing only in one local area. Although the greatest number of threatened species are in tropical forests, most ecological zones are undergoing similar habitat destruction. To take an example close to home for many in the United States, the decline of songbird species in North America has been thought to be due to deforestation of their winter grounds in Central America. Careful long-term research on two warbler species has shown that their decline is due instead to changes in their summer habitats in the Northeastern United States. The causes are suburbanization and fragmentation of forests into housing lots, bringing the predators and competing species who live on the human periphery (Holmes and Sherry 1992). The loss of songbirds is not casually or immediately observed—it requires counting gradual population changes of multiple species over wide areas and over long periods of time.

The total number of species on the planet is unknown. Biological systematists Peter H. Raven and Edward O. Wilson point out that only 1.4 million species are known and classified, of a total estimated from 10 to 100 million species alive. With at least 0.5 percent of known species going extinct annually, even to catalog existing species, much less preserve them, is impossible: "the few hundred systematists available are woefully inadequate to complete the task while most of the species are still in existence" (Raven and Wilson 1992, 1100). They propose a more modest goal: "to chart the outlines of global biodiversity, use it for humanity's benefit, understand it scientifically, and preserve an intelligently selected sample of it for the future."

In a sense, humanity's primary reaction to species loss so far has been unplanned adaptation. We have psychologically accepted the loss of those creatures. We have also adapted by substituting human and technical means for the functions they perform, for example, hunting deer to control populations once limited by wolves.

American law (the Endangered Species Act) and American public thinking tends to focus on preserving individual species rather than their habitats. The species focus has been criticized as leading to a one-species-at-a-time approach that is inefficient and expensive, resulting in extinctions before the legal mechanisms can catch up. There is a growing acknowledgment of this problem and corresponding efforts to set aside habitats, whether American wetlands or Amazonian rainforests. But land is valuable. Development and resource-extraction industries are usually much more politically potent than the call for an abstract entity such as habitats, making it politically difficult even for a wealthy country to make such commitments.

Ozone Depletion

To understand both ozone depletion and the greenhouse effect, we must start with the sun. The part of the sun's light that humans can see is composed of a spectrum of colors, as can be seen in a rainbow. In addition to this visible light, the sun gives off other forms of light or radiation not visible to our eyes. On either side of the sun's visible light spectrum are *ultraviolet light* (UV) and *infrared light,* also called radiant heat. Ozone depletion involves ultraviolet and the greenhouse effect involves visible light and infrared. The two pertinent facts about these forms of light are: (1) ultraviolet, if unfiltered, is very damaging to biological tissue, and (2) visible light striking an object warms it up, and the warm object then reradiates infrared. We can summarize the essence of each of these two problems as follows: ozone depletion lets more ultraviolet through the atmosphere to hit us. The greenhouse effect warms the planet because the atmosphere passes visible light through to the earth's surface while blocking some of the resulting heat from getting away. These two effects are illustrated in figures 2.1 and 2.2.

Ozone (O_3), a gas molecule made up of three oxygen atoms, is formed naturally in the stratosphere—the earth's upper atmosphere. This stratospheric ozone, along with clouds and particles, shields humans and other species from the sun's ultraviolet radiation.[4] The ozone layer is being thinned by human-made chemicals called chlorofluorocarbons, or CFCs, (used for refrigerators, air conditioners, foam insulation, and many other things) and to a lesser extent by halons (used primarily in fire extinguishers). These human-made chemicals released on earth gradually rise to the stratosphere where they react with ozone and convert it into other substances. Ozone loss has been most pronounced in Antarctica, because the chemical reactions occur more quickly in cold air, but ozone loss and corresponding increases in UV have now been measured in the populated midlatitudes (Appenzeller 1991; Kerr and McElroy 1993). Unfortunately, CFCs take ten years to reach the stratosphere, and they remain there from fifty to one hundred years. Even if there were an immediate and total ban on CFCs, ozone loss would worsen for a decade and persist past our lifetimes, though it would recover in approximately one century.

Higher UV levels would increase human skin cancers worldwide and damage other species. UV also causes eye cataracts and reduces human immunosuppression, increasing the rate and severity of diseases. The U.S. Environmental Protection Agency (EPA) estimates that the elimi-

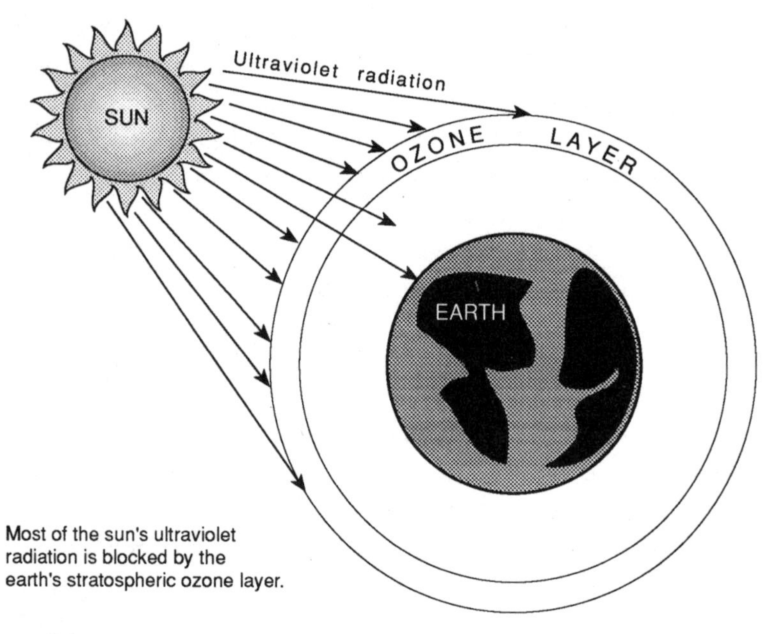

Figure 2.1
The ozone layer prevents harmful ultraviolet light from reaching earth's surface.

nation of CFCs by an international treaty (described below) will prevent 27 million deaths that otherwise would have occurred from skin cancer during the next century.[5] For comparison, 15 to 20 million people lost their lives in World War I and 40 to 50 million died in World War II.[6] The numbers are approximations but make the point that the threat of human fatalities from ozone depletion had been about that of a major world war.

Many species are damaged by UV, an indirect threat to humanity that many scientists feel is more serious than the direct medical effects on humans (Bundestag 1989). Phytoplankton are of particular concern because they are the basis of the entire oceanic food chain. The present loss of Antarctic ozone has been shown already to be reducing phytoplankton productivity by 6 to 12 percent (Smith et al. 1992).

The problem of ozone depletion has entered the political realm several times. One important cause of the early U.S. government ban on CFCs in spray cans was pressure from individuals who, as consumers, rejected

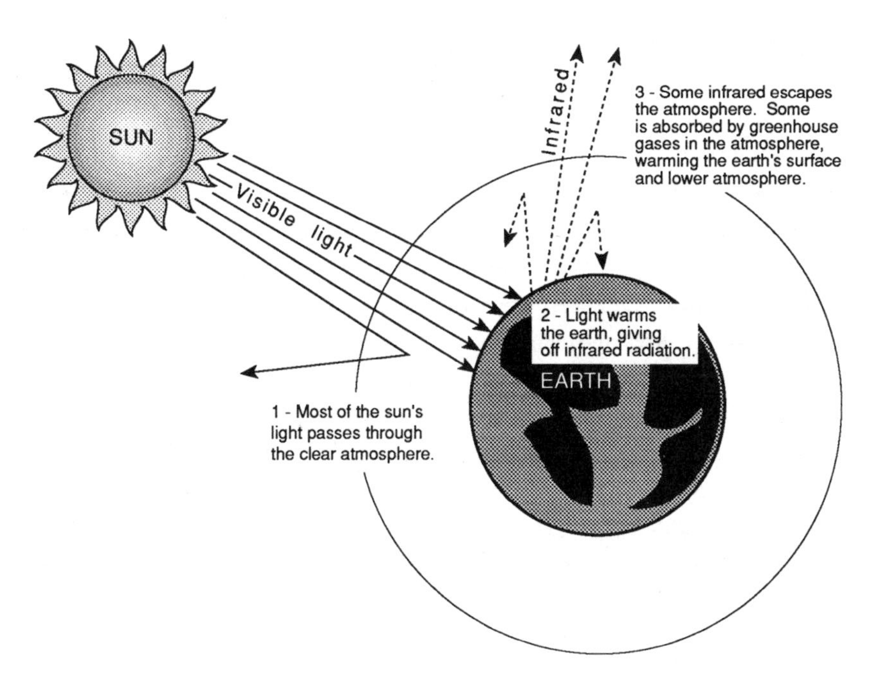

Figure 2.2
The greenhouse effect absorbs infrared (heat) and keeps the earth warmer.

spray cans and, as voters, wrote to their representatives (Benedick 1991; Kempton 1993). Later, as the United States was preparing to enter into binding international prohibitions on CFC production, the U.S. Secretary of the Interior at the time, Donald Hodel, proposed that the government did not need to ban CFCs. Instead, he suggested that industry could continue to produce them and individuals could adapt by using sun hats, lotion, and sunglasses when outdoors.[7] This is an example of an adaptation proposal, although not a very well considered one. At the time, it was perceived as so naive about the worldwide diversity of those who work outdoors and the effects of UV on other species that it became grist for editorial writers and cartoonists and was quickly disowned by the Reagan administration. The public uproar over Secretary Hodel's suggestion illustrates one case in which society was given a clear choice and resoundingly chose prevention over adaptation.

Ozone depletion has recently entered the political domain again, with a political commentator claiming that CFCs do not even significantly deplete stratospheric ozone (Limbaugh 1993), a claim that seems wildly off base to scientists working in this area, given the clear scientific evidence (Taubes 1993).

In summary, ozone depletion is a very serious problem. If it had been allowed to continue unchecked, during the next century worldwide human casualties directly due to UV would have been approximately equal to those of a twentieth-century world war. Fortunately, much progress has been made. The response to ozone depletion represents a great success story, diplomatically and scientifically. International agreements to eliminate production of CFCs and halons have been signed and are being implemented in the United States and globally (Benedick 1991). A variety of alternative substances have been developed for virtually all their functions—some are more expensive or do not work as well, others are cheaper and better. The only remaining debates are on secondary issues: whether to accelerate the phase-out schedule, how long to allow use of certain CFC substitutes that cause lesser but not zero damage to the ozone layer, and what to do with the CFCs still in circulation in existing refrigerators and air conditioners.

Global Warming

The earth's atmosphere creates a natural greenhouse effect that, like a greenhouse building for plants, traps the sun's heat and keeps it warmer inside.[8] Visible light passes through the atmosphere, striking the earth's surface and heating it. This surface heat radiates back up as infrared light. Naturally occurring greenhouse gases absorb part of the infrared in the lower atmosphere, thus keeping the earth warmer. Without that natural greenhouse effect, the earth's average surface temperature would be a chilly 0°F (−18°C) rather than its present global average of 59°F (+15°C). The greenhouse effect is illustrated in figure 2.2.

The primary natural greenhouse gases are water vapor (in clouds), carbon dioxide (CO_2), methane, nitrous oxide, tropospheric (near-surface) ozone, and other chemicals having a lesser effect. Human activities are increasing the concentrations of all these greenhouse gases except water vapor. Humans have also introduced new greenhouse gases, CFCs and other halocarbons, that do not occur naturally and cause ozone depletion as well as greenhouse warming. Altogether, these increased greenhouse gases are expected to alter climate globally. Even though these gases occur in trace quantities (e.g., carbon dioxide is 0.03% of the atmosphere), their increases can be unambiguously measured. However, scientists disagree as to whether any climatic effects of the increases are yet definitively measurable. For example, some scientists who analyze temperature records from weather stations conclude that global average temperature has risen, whereas those analyzing temperature records from satellites see no definitive change yet.

The greenhouse effect is a natural phenomenon. Anthropogenic (human-caused) increases in the greenhouse effect are called *global warming* or *global climate change.* The term *global warming* is more common in analytic literature, and we usually use it in this book.

The major anthropogenic greenhouse gases are shown in table 2.1. The gases are ranked by their effect on increasing radiative forcing, that is, proportional to the amount they change the heat balance illustrated in figure 2.2. The effect column in the table takes into consideration both the potency of each gas and how much of it humans released during the eighties (Office of Technology Assessment (OTA) 1991, 45; Caldeira and Kasting 1993). CO_2 alone accounts for 55 percent of the anthropogenic effect. Since a worldwide CFC ban is already being phased in,

Table 2.1
Major anthropogenic greenhouse gases

Greenhouse gas	Effect	Lifetime (Years)	Activities causing emissions
Carbon dioxide (CO_2)	55%	1000	primarily fossil fuel burning, secondarily deforestation
Chlorofluorocarbons (CFCs)	24%	100	refrigerators, foam insulation and packing, previously spray cans—now banned
Methane (CH_4)	15%	10	rice paddies, ruminant animals, coal mining, natural gas leaks, landfills, and biomass burning
Nitrous oxide (N_2O)	6%	150	nitrogenous fertilizers, burning fossil fuels and biomass

and methane and nitrous oxide together are only 21 percent, the table makes clear that CO_2 is the most important greenhouse gas in its total effects. It is also more important because its lifetime in the atmosphere is much longer than that of other anthropogenic greenhouse gases—over 1000 years of significant effects, as compared to 10 to 150 years for the others (Caldeira and Kasting 1993). We include table 2.1 and the subsequent one in order to make such overall comparisons, not for the exact percentages, most of which are subject to some debate.[9]

Another way to apportion causes for the increased greenhouse effect is by human activity. For example, when sorted by activity rather than by gas, CO_2 from energy is combined with methane leaks from natural gas pipelines, but is separated from CO_2 from deforestation. Gross categories of activities are shown in table 2.2. Land use modification includes tropical deforestation. Tropical deforestation can be seen to be only a small source of greenhouse gases in comparison to energy, although it is still a major concern for species loss, as explained previously.

It may be helpful to bring abstract global numbers to human-scale terms. For example, driving a short distance to a shopping center and burning one gallon of gasoline in the process will produce 22 pounds of CO_2. This would be enough to double the CO_2 content of a half-million cubic feet of air, about the volume of air contained in the entire shopping center (DeCicco et al. 1990, 11). And that's only one shopper on one trip! Multiply by the number of gallons used by one car in a year and by the number of cars on the earth, then add in all other fuel uses. This

Table 2.2
Major human activities increasing the greenhouse effect, adapted from Lashof and Tirpak (1991)

Human activity	Effect
Energy use and production	57%
Industrial (nonenergy, mainly CFCs)[1]	20%
Agriculture	14%
Land-use modification	9%

Note: The role of CFCs is estimated to be somewhat smaller in this table than in table 2.1, partly due to different assumptions about their effects on climate.

illustrates how our seemingly insignificant human activities can change the whole planet.

From figure 2.2, it would seem that the main effect of increasing greenhouse gases would be to raise the earth's temperature. The reality is more complex: since temperature drives evaporation and the movements of the atmosphere and oceans, the dynamic effects on weather are more important than simple temperature increases. Using sophisticated computer programs, climatologists have tried to predict the effects of increased CO_2 and other greenhouse gases.[10] Commonly predicted effects include dryer weather in midcontinent areas (including the U.S. midwest), sea level rise, more violent storms, and northward migration of climate-sensitive species (Intergovernmental Panel on Climate Change 1990; Abrahamson 1992; Peters and Lovejoy 1992). Small temperature changes have surprisingly large effects on climate. For example, past ic ages were caused by average temperature reductions of approximatel 9°F (5°C). However, all the specific effects are predicted from compl cated computer models, which vary somewhat in results even when giv the same input assumptions (Cess et al. 1993). Although most climat ogists consider the above-listed effects plausible, they cannot be predict with certainty.

If effects like these do occur, human societies would have to adapt such consequences as being unable to grow some crops where they now grown, more frequent and more violent hurricanes, flooding coastal areas, and northward movement of vector-borne tropical eases. Other living species would have to cope with unusually r ecological changes, requiring migration to survive. Many species w not be able to migrate quickly enough and would become extinct.

In addition to the uncertainty inherent in computer simulatio second problem is the level of CO_2 used in the simulations. Clima gists have been modeling based on a benchmark of doubling of over 50 to 100 years. However, more recent calculations show that CO_2 will be closer to an increase of six times (Sundquist 1990) o times (Cline 1992) and will occur in 200 to 300 years. These levels of CO_2 appear to have much more severe effects on huma the environment than doubling would (Manabe and Stouffer 199 occasionally refer to the Sundquist figure of sixfold increase as a

tual or peak CO_2 level. It is very approximate, but it is likely to be closer to the actual number than doubling, which is a scientific benchmark and was never intended to be a prediction.

The nature of the problem also makes global warming difficult as a political and diplomatic issue: the effects are not known with certainty, the worst effects will occur far in the future, and the greenhouse gases causing the change are central to many industries. As with the problem of ozone depletion, negotiations have begun on international treaties to address global warming (Grubb et al. 1993). However, many familiar with both the ozone and global warming negotiations predict that the latter will be more difficult. For example, in the judgment of one of the key diplomatic figures in the ozone accords, "Due to scientific uncertainty and the complex issue of economic feasibility, one can safely predict that it will take a much longer time to get this new [global warming] regime working. . . . Partial achievements may be at hand in the years to come. However, definitively arresting global warming will remain one of the major tasks of the twenty-first century" (Lang 1990/1991, 174).

How could global warming be arrested? Policy approaches include adaptation and prevention and an additional approach called geoengineering. These can be combined, but to simplify the following discussion they will be considered as mutually exclusive policies.

The adaptation strategy is that, rather than trying to prevent global warming, we would switch crops, move farmers north, and build dikes to hold back a rising sea. Serious economic analyses have argued that such an adaption strategy would be less expensive than reducing the quantities of greenhouse gases. These economic analyses have been criticized, however, because they assume an independence of economic activity from nature, typically assign a monetary value of zero to environmental damage, discount future costs to their present value, and do not include damages to other countries.[11]

The geoengineering approach to climate change, like adaptation, would not change greenhouse gas emissions. Rather, this approach would use planet-scale engineering projects to nullify an increased greenhouse effect. Proposals have included spreading dust in the upper atmosphere to reduce incident sunlight and seeding the ocean with iron

to increase the growth of phytoplankton, which take in CO_2. A National Academy of Sciences (NAS) panel found some geoengineering options "feasible" with existing technical means, at "reasonable cost" (1991a). However the NAS panel cautions that the earth's systems are nonlinear, involve complicated feedback, and are likely to be unstable and exhibit unintended side effects. Other scientists, who seem to be appalled by geoengineering solutions, make statements mixing biology and morality such as "you don't tinker with a perfectly healthy ecosystem to clean up humanity's mess" (Roberts 1991), or arguing that it is wrong for us to change the global environment in ways that require constant human maintenance for the future.

The prevention approach is to reduce the emission of greenhouse gases. Since there is currently no technology to remove the CO_2 from exhaust gases, most prevention schemes limit the activities that emit greenhouse gases. Some reductions in CO_2 could be accomplished by increasing energy efficiency and would probably be profitable, in that energy cost savings would exceed the cost of changes to equipment or operations. Emissions would not need to be completely eliminated. One estimate is that stabilizing climate change would require reductions of 15 percent in methane and 60 percent to 80 percent in CO_2 and the other major greenhouse gases (Lashof and Tirpak 1991; IPCC 1990, xviii). Fossil fuels are crucial to transportation, industry, and home energy. Reductions of 60 to 80 percent may seem impossible to many people if we are to maintain our standard of living.

Analyses based upon aggregate economic data have consistently argued that major changes in the energy system will be unacceptably expensive (Lave 1981; Manne and Richels 1989; Nordhaus 1991). A different picture emerges from analyses examining specific technological substitutions. For example, Johansson and colleagues have recently analyzed renewable energy, technology by technology, estimating that deployment of cost-effective renewables and energy efficiency could, by 2050, reduce world CO_2 emissions by about 40 percent per capita (Johansson et al. 1993, 6). Although such per capita improvements are impressive, after factoring in expected world population growth, global CO_2 emissions would be reduced by only 26 percent by 2050; developing country emissions would increase slightly while industrialized countries would reduce CO_2 emissions by two-thirds (Johansson et al. 1993, 4–6). The 26 percent figure is well short of the 60 to 80 percent reduction needed for climate stabilization. This analysis conservatively assumed that renewables would not be used if they cost more than fossil fuels and that no credit would be given for environmental, economic development, or energy-independence benefits of high renewables use. In contrast to the aggregate economic analyses, it predicts a large, cost-effective resource of non-CO_2 technology. Thus, although this study is more optimistic than those based on aggregate economic data, it assumes adoption of only renewables cheaper than current fuels, so population growth and third-world development limit total CO_2 reductions.

Despite other differences, the above studies both lead to the conclusion that major climate-change prevention efforts will not occur on their own—they will require concerted changes in law and policy. Some advocate broader structural changes to address environmental problems (Schnaiberg 1994). A recent study by the U.S. Office of Technology Assessment (OTA) itemized some policies that could reduce CO_2 emissions: taxes on energy or carbon-emitting fuels, tax credits or low-interest loans to encourage renewable energy sources and energy efficiency, standards to require new products (cars, appliances, buildings) to achieve higher energy efficiency, federal support for research and development of new technologies, and public information and education (OTA 1991, 13–17). The OTA points out that many of these CO_2 reduction policies would have additional benefits, such as reducing local air pollution and increasing energy security.

3

Cultural Models of Nature

We begin our examination of how laypeople view environmental problems by describing the cultural models Americans use to understand nature and humanity's interaction with it[1]. The models of nature described here apply to any environmental problem, not just global ones, and subsequent chapters will demonstrate that these general models are the basic conceptual underpinning of popular American thinking about the environment. They are used to understand global environmental problems, they reenforce and justify environmental values, and they are the basis for reasoning that leads to preferences for some environmental policies over others.

When we began this study, we did not anticipate cultural models that would so broadly cover the ecosystem and humanity's relation to it. We were looking more specifically for models of global warming. However, following anthropological interviewing guidelines, we preceded our specific interview questions with general ones such as "*Would you say that protecting the environment is important? Why?*" and "*Would you say that you have environmental values?*"[2] These questions, as well as specific questions asked later in the interviews, were answered in ways that made sense only after we hypothesized the broad models about nature and humanity that we describe in this chapter. We support our inferences with quotations from those interviews, letting our informants speak in their own words, and with our survey results to illustrate the frequency of these models across the population.

This chapter covers three sets of general environmental models we discovered. First are models concerning nature as a limited resource upon which humans rely. Second is the pivotal cultural model of nature as balanced and interdependent, with the derivative models of "chain reactions" that potentially can ripple across species, and of the unpredictability of such interdependencies. Third are the cultural models of society and nature: the market's devaluation of nature, the separation from nature that leads to failure to appreciate it, and the American idealization of the environmentalism of primitive peoples. For the second and third sets of models, we compare the cultural models of laypeople with the models of scientists.

Model of Human Reliance on a Limited World

The models described here involve the related ideas that humans are part of the environment and depend upon it, that the planet is limited in size, and that our wastes do not disappear but enter cycles and eventually return to us.

One of the most commonly expressed justifications for protecting the environment is that it is a fundamental basis for human life. For example:

> [Protecting the environment is important] to a certain extent because we live in it. . . . We have to breathe it, we have to live with it . . . We're mixed in with it.—Frank (building contractor, logger)

In the most survival-oriented form, human dependence on the environment is expressed as a health concern. For example, one person was worried about "breathing chemicals" from pollution, another noted difficulty breathing on days of poor air quality, and a third claimed that many diseases "probably originate from human-created problems with nature." Using a metaphor from drug use, another person said that damaging the environment was "self-destructive, the same way that taking heroin is."

Informants often used examples of pollution from nearby sources that could readily be seen. Several furthermore saw the entire planet as a closed system, as revealed in statements such as:

> We're all on wells here. . . . It's common sense not to dump it [toxic paint waste] in the ground. . . . When you're dumping it a hundred, two hundred feet from your well, that's one thing. But when you're pushing that in the ocean, you can

ignore it for a few years. . . . the whole issue is just one of time.—Nick (small manufacturing plant owner)

Similarly, Wilbur (a retired fireman) asks "How much can you dump? As big as the ocean is, it's got to come back somehow." Doug, a research scientist at a pharmaceutical firm, expresses the problem with a striking metaphor from his work.

One of the interesting stories I've always gone by is the simple bacteria, in a . . . colony. If you put a bacteria in a certain medium and let it grow, it starts building, but what happens—it starts creating waste, and as it starts growing and growing it starts making more waste and more waste. Well, eventually it ends up dying in its own waste. You know . . . if I can look at a simple bacterial colony as the way I look at the earth, I can actually see that possibly happening.—Doug (pharmaceutical scientist)

The frequency of such statements in the semistructured interviews suggested that this model is widespread, an inference supported by national probability samples. For example, 80 percent of a U.S. sample agreed with the statement "The earth is like a spaceship with only limited room and resources." (Dunlap, Gallup, and Gallup 1993). Similarly, in a question testing the idea that earth's ability to support humans is limited, 61 percent *disagreed* with the statement "The earth can support a much larger world population than exists today." In response to a related question in a much earlier survey in Washington State (Dunlap and Van Liere 1978, 13), 96 percent agreed "Humans must live in harmony with nature in order to survive." Although the results of our semistructured interviews might suggest some rewording of the questions used in those surveys, the probability samples of those surveys demonstrate the wide distribution of the cultural model that the earth is a closed system upon which humans depend.[3]

For many, the concept of dependence goes far beyond health to ideas that are expressed by using the metaphor of "home." (These ideas are consistent with the derivation of the word *ecology,* from the Greek word for home.) Statements made in our interviews include "The environment is our home." (Kate, college student), and "When you destroy your environment, it's like burning down your home." (Walt, retired machinist). One person expresses this more literally than metaphorically, saying that our planet is fragile and unique:

How many places do we have to go once we destroy this one, you know? I think it's kind of primary. . . . There's not many places to escape to. We have to deal with what we have here and do the best we can to protect it.—Jenny (social studies teacher)

These quotes reflect a reversal of the traditional view of home as a shelter from a vast and threatening nature. For whatever cultural reasons—whether the view of earth from the moon, increased public environmental awareness, or other factors—"home" has expanded to the environment or the entire planet.

Informants sometimes included humanity's dependence on the environment as involving psychosocial health, as well as physical health. Paige argues for preservation of green areas in the inner city.

Some things in the environment you just need, and if it's not there for your need, you're going to suffer. . . . you need to feel grass, to see the greenery out here, somehow you need it to keep this thing [points to head] going. You know, it's like a cement jungle. I mean, inner-city Trenton, they're crazy. I mean, if they've got one blade of grass in front of their house, they're doing good. . . . I think it makes them, you know, violent. . . . It's like a concrete jungle out there. And, I mean, it's not that it's a ghetto . . . they don't see anything pleasant, and therefore they're not going to be too pleasant themselves.—Paige (manufacturing worker)

A similar psychological need is cited by an elderly woman who has personally observed local loss of species due to housing development in her neighborhood: "You need birds around. You can hear 'em singing, it makes you feel better" (Jane, retired insurance actuary).

Several statements in our survey support these models of human dependance on the environment. We cite just one statement here, which contrasted human dependence on the environment with dependence on the economy.

21 We should be more concerned about the environment than the economy because if the environment is all right we can at least survive, even if the economic system is not in good shape.

Earth First!	Sierra Club	Public	Dry cleaners	Sawmill workers
97	85	73	67	59

(Recall that the numbers are percentages agreeing with the statement in each survey group.) Agreement varies across the five survey groups in a predictable pattern. Nevertheless, majorities of all groups agree with it, even those who have been economically hurt by environmental regulations.

In sum, there is a broadly shared recognition that humans fundamentally depend on the environment and that the earth is a closed system in which our effluents eventually return to us. The natural world is described metaphorically as a home and literally as a limited resource meeting physical and psychological human needs. One practical significance of this cultural model is that it provides a strong utilitarian motivation for protecting the environment—a motivation seen as common sense, almost precluding direct counterarguments.

Models of Nature as Interdependent, Balanced, and Unpredictable

In addition to the cultural model of nature as limited and humans as dependent on it, there is a subtler set of models about interactions within nature. As these emerged again and again from many parts of our interviews, we began to see them as among the most important and most central set of American cultural models of the environment. Three interrelated concepts are involved. First is a model that different parts of nature, for example, different species or ecological conditions, are so interdependent that changing one can have multiply linked chain reactions on a series of others. Second, these interdependencies are so complex that the interactions are impossible for humans to predict in advance. Third is a resulting proscription against human interference with nature. As we shall see, these ideas are related to ideas of species interdependency in scientific ecology, but their popular versions take on a modified form and are applied more broadly. A surprising result of our interviews was in finding a wide variety of people, across the range of educational levels, articulating these concepts. We suspect that they are more widespread among the general public than other scientific principles of comparable complexity.

Interdependencies and Chain Reactions

Most of our interviewees had a clear model of interdependencies in nature. They expect that perturbations, like removing or adding a species or changing climate, will cause other significant changes. Many informants refer to these interactions using the term *chain* or *chain reaction,* lay terminology that seems to derive both from the physics terminology for nuclear reactions and from the biological term *food chain.* The term *balance* is also used frequently in this context, presumably referring to a *balance of nature.*

Even though none of our questions asked about such interrelationships, informants mentioned them frequently in order to answer other questions, as in the following examples:

Warmer climate would lead into a rise in your sea level. And . . . that will lead into plant and animal extinction. . . . Life has a chain, and when they destroy insects, they destroy bird life, and destroying bird life destroys other life. That's a chain reaction, for all of that . . . from the smallest microscopic life [on up] . . .—Walt (retired machinist)

[If there are species extinctions] that would have a rippling effect. . . . These birds are destroyed, and they eat these insects, and insects proliferate, and more damage [occurs] a thousand miles away.—Charles (coal mine construction worker)

Occasionally, the model of species interdependency was explicitly tied to an existing balance of species.

[All animals] are here for a reason. They gotta be food for something. . . . If they're food for a certain animal, and they become extinct then [that animal] will try to get something else. . . . It'll unbalance everything. . . . Something happens, one dies, which makes something else die, which in turn makes something else die. [A species extinction] might cause a chain reaction.—Cindy (housewife)

Cindy's statement that all animals "are here for a reason" gives the interdependency of species a theological or teleological explanation. Others simply described the balance, stating the way they believed nature works.

Some informants consider the balance to be quite delicate. For example, an educated, politically active mine worker, Emma, reacted as follows to an interview question suggesting that we let climate change proceed and adapt to it rather than trying to prevent it:

Whoever thinks that has somehow escaped the logic of exponential change, that you can set off a chain reaction that just keeps getting bigger and bigger as it goes down through nature. Whoever says that sees changes as still kind of linear. You throw the ball and it's only going to go so far. And I think that is largely the policy that is guiding our establishment right now. . . . And I think that it's foolhardy.—Emma (coal loading machine operator)

Emma's claim of "exponential change" would literally apply to the concept of chain reactions in physics, in which the effect (in that case, the number of nuclei giving off neutrons) becomes exponentially larger with each successive generation in the nuclear reaction.

Whereas nature is seen as fragile in the face of large perturbations such as species extinctions or climate change, it is seen as resilient to small changes, even having a self-healing capability.

We, all of us, have built our camps and so forth too near the water. We didn't realize. . . . We know now that things should be back so that the water can filter and clear. . . . It's too late to do too much about that, but we can look down the road and be sure that not anymore's done, see, because the earth healing itself can heal it up to a point, and then it's going to need help.—Catherine (retired science teacher, environmentalist)

The self-correcting model was also used to oppose the idea of deliberately trying to counteract one human change by creating an opposite one. In particular, this was mentioned in response to our interview's proposal to cool climate in order to neutralize the anthropogenic greenhouse effect.

If we tried to counteract that, we could be in an even worse situation. . . . [We should do] nothing by man other than reverse our habits, but not reverse what's happening. . . . Hold off on everything we've done to cause the greenhouse effect, and see if it can self-correct or see if we can live with the situation as it stands.—Kate (college student)

These quotations suggest a belief that nature will adjust to small changes, yet is vulnerable to large ones. The self-correcting component of the balance of nature cultural model seems limited: if perturbations are small, nature will right itself, but if they are too large it cannot.

The significance of this model of interdependencies and chain reactions is that it leads Americans to be conservative about changing nature, even parts of nature that are unimportant to us. In this sense it greatly extends the first model of a limited environment by making all aspects of that environment potentially relevant to humans. It leaves undefined which parts will in fact be relevant and how large a change will cause chain reactions.

Interdependency of Nature Prescribes Nonintervention

Our interviews show that informants augment the belief that natural systems are complexly interrelated with the belief that humanity cannot understand these interrelationships fully. A widely held opinion that humans are unable to predict these interactions leads them to a prescription to avoid human disturbance. We call this linked set the *nonintervention model.*

For example, an informant reacts as follows to the interviewer's description of human-induced climate change:

> [With global warming] everything will just get all out of whack. I think once you start mucking with something that big it will chain react through everything that we're used to in unpredictable ways.—Pervis (coal mine wireman)

When asked about the wisdom of using technology that could change climate for the better, James says we would probably "foul it up," because that would be "fooling with nature," and that it is "pretty tough for man to duplicate nature." Both Pervis and James invoke the model of nature as interdependent and complex, which leads them to advise against human intervention.

A conservative congressional staff member elaborates this point further and links it to a similar tenet (apparently drawn from Karl Popper) of conservative social thought. He begins in response to our request to describe *the relationship between human society and nature.*

> It seems that things happen for a reason. . . . It all seems so precise and calculated that . . . you want to go very lightly into a situation where you're going to be tampering too profoundly with the environment. . . . I'm a conservative, obviously, and one rule of thumb that conservatives live by is the law of unintended consequences: that if you try to engineer human behavior, things are going to happen that you never anticipated. And I think the same could be true of the environment.—Gerard (legislative counsel)

Another congressional staff member, although he is from the opposite end of the political spectrum, invokes the same model to argue against large-scale geoengineering to fix the greenhouse effect.

> Some scientists are suggesting that you could put 500 tons of iron into the ocean to make algae bloom. The algae would soak up carbon. It's obviously bullshit. On the face of it, it's ridiculous. *Why?* You're already conducting one large experiment with global warming. Now they're suggesting conducting another large-scale experiment. Even if that solves the problem, they don't know what

the other ramifications are. It's offered as a panacea. That's Ronald Reagan thinking.—Alvin (legislative aide)

The quotes in this section show that the nonintervention model is based on a two-part model covering both the complexity of ecosystem interactions and the limits of human knowledge and prediction. This line of argument is not limited to highly educated professionals or linked to any particular political philosophy.

Although most informants in the semistructured interviews used a nonintervention model, two or three did not and generally expressed less concern about human impact on the environment. For example, Ronald feels that humans would know if the environment were seriously threatened and would act to avert any dangers.

I think that . . . the mind of man is superior to nature and that if man feels that he has extended nature or is threatened because of his overuse of nature, that . . . he has the capacity—spiritually, and mentally, and emotionally—to come up with some solution. I don't think he'll just engage in some sort of long-term suicide.—Ronald (resort proprietor)

A congressional staff member also rebuts the nonintervention model. He summarizes it concisely but seems to mistakenly believe that this model is limited to the "environmental community."

See, the environmental community would never have ever considered as a resolution of the global warming issue, bioengineering the plankton to absorb the CO_2. Okay. Because that is an active intervention. . . . I think the theory goes, or the thesis goes that we don't know enough about the environment to positively intervene to resolve issues. So that the best way to do it is to just leave it. And to try to leave the environment as it was without us there . . . the theoretical reason is that it's too complicated. The natural system is too complicated to replicate or to positively intervene in. . . . We don't know enough. We'll never know enough. It should not be done.—Luke (congressional staff)

When asked for his own position, Luke goes on to criticize the nonintervention model.

My position is, that, in fact, we probably are going to make this transition to the point at which it would be almost impossible . . . without positive intervention. . . . Let's say everybody [in the world] is at the same level of economic development that we are, and consuming anywhere near, and emitting anywhere near what we do right now. The concept of just sort of letting natural forces handle it would be very, very difficult to do, given that type of situation. So you actually have to actively go and try to do it.—Luke (congressional staff)

Whether right or wrong, Luke and Ronald were very unusual among our informants in thinking that humanity could actively manage nature.

Paralleling the fragility-in-the-large, resilience-in-the-small concept of nature, the nonintervention model seems dependent on several factors: the scale of the change, its unpredictability, and its uncontrollability or irreversibility. The five informants who discuss hunting all believe that animal populations can be managed, or limited, in a predictable way by socially regulated human intervention (hunting, in particular). Usually this is seen as good for the animals, to prevent overpopulation and consequent starvation, disease, and unpleasant death.

> I think you have to have a balance [in the animal population]. If you don't you can get an overabundance of them. So there again, you've got to equalize, try to keep it within a safe number . . . take a certain amount you kill and make sure there's [a] certain amount that are living. . . . To me, life all the way through has to be an equal situation. You make sure that you have enough supply for years to come and yet enough for . . . the present.—James (farmer, custodian)

> I get awfully upset to think there's a moose hunt because they're nothing but just like tame cattle. But they say there's so many of them that if they aren't killed and eaten . . . then they're going to get diseased, and they're going to die. . . . A controlled atmosphere is right. I think of control to be sure to take care of the animals. [If] there's too many or if they're diseased or something, they should be weeded out.—Catherine (retired science teacher, environmentalist)

These informants who advocate hunting to control population do not advocate human manipulation of nature in general. James, for example, was quoted earlier advising against "fooling with nature." Instead, the hunting examples seem to be interventions considered small enough or limited enough that they can be predicted and controlled successfully by humans.[4] A similar exception must be applied to an intervention like agriculture, which massively changes nature but does not seem to generate concerns about interdependencies. (Cronon (1991) suggests that human disturbances such as agriculture and mining become so familiar that they seem more like second nature than human disturbance).

Only a couple of informants qualified the nonintervention model, saying that other things, such as economic well-being, are as important to humans as the relationship with nature. Gerard, for instance, said that environmentalists were saying "don't develop," but that they should understand that other countries want to grow the way the United States has. Ronald made an argument for fairness very similar to that of Tucker (1982).

I'm not sure that I would agree with those that simply say we should leave . . . vast tracts of wilderness. I don't [think] that we should leave these in as pristine a state as we can in order for a few hikers to go through them occasionally, you know, and take pictures of them, and get a big thrill. I don't quite understand that. I mean, it might be very nice to leave vast tracts of the environment untouched or relatively untouched on the face of this earth, but basically it doesn't make sense if it's just going to be for the benefit of a few people.—Ronald (resort proprietor)

This statement implies that the primary goal of wilderness conservation is human benefit. We will discuss this and opposing views in the chapter on values.

In short, the nonintervention model holds that nature is interdependent, that its relationships are so complex as to be unpredictable, and that therefore human modifications are unsafe. Based on these findings from the semistructured interviews, we added related questions to our fixed-form survey. The survey results confirm that these models are broadly shared and fill in some of their elements.

105 Nature has complex interdependencies. Any human meddling will cause a chain reaction with unanticipated effects.

Earth First!	Sierra Club	Public	Dry cleaners	Sawmill workers
97	89	77	76	63

109 Nature may be resilient, but it can only absorb so much damage.

Earth First!	Sierra Club	Public	Dry cleaners	Sawmill workers
94	93	97	93	85

Agreement dropped precipitously when the resilience argument was used to justify a lower level of environmental protection. Even when we use loaded language ("radical measures"), no group came even close to majority agreement.

29 The environment may have been abused, but it has tremendous recuperative powers. The radical measures being taken to protect the environment are not necessary and will cause too much economic harm.

Earth First!	Sierra Club	Public	Dry cleaners	Sawmill workers
0	7	23	17	33

The survey also makes it clear that this principle of minimizing human intervention is broadly applied to global climate change.

57 Global climate change would disturb the whole chain of life.

Earth First!	Sierra Club	Public	Dry cleaners	Sawmill workers
100	85	93	90	81

126 Global climate change would be bad even if it didn't cause humans any harm, because it is not a natural change.

Earth First!	Sierra Club	Public	Dry cleaners	Sawmill workers
94	74	87	83	67

Although there is majority agreement across groups, these questions also reveal patterns of variation across the spectrum and occasional blips up or down. These patterns of agreement and disagreement across groups are analyzed systematically in chapter 8.

As mentioned in chapter 2, most surveys have not addressed any of these questions. One survey that does (Dunlap, Gallup, and Gallup 1993) found that 87 percent of a U.S. national sample agreed that "The balance of nature is very delicate and easily upset." Another of their questions touched on part of the nonintervention model, although it unfortunately did not specify complexity and unpredictability as causes: 66 percent of the U.S. sample disagreed with the statement "Modifying the environment for human use seldom causes serious problems." In short, our own survey, as well as the limited national polling that exists on this topic, support our inferences of the cultural model of complex interdependencies, unpredictability, and the prescription of nonintervention.[5]

Cultural Models versus Scientists' Models of Ecology

How does the cultural model of interdependencies and unpredictability relate to that of scientists? We compare our lay informants' cultural models to those of ecologists, that is, scientists who study ecosystems. As we compare the two, we recall the example of Tukano beliefs about fish spirits, a cultural model vastly different from a scientific one, yet a model that guides people to manage their environment effectively.

Most ecologists would add substantial qualifications to the cultural model. For example, they are uncomfortable with an emphasis on a balance of nature, they qualify our informants' model of fragile interdependency and unpredictability, and they do not use the term *chain reactions*. We show in this section how the cultural models differ from the ecologists' models, then argue that the cultural models in this area—unlike some models discussed in the next chapter—are nevertheless reasonable simplifications.

It could be argued that the American cultural model of balance of nature draws from an older scientific ecology that is now in disfavor. The lay cultural model of the balance of nature parallels ideas developed by early ecological science: a stable climax stage of ecosystems (Clements 1916, cited in Worster 1977) and homeostasis, equilibrium and balanced ecosystems (e.g., Odum 1969). However, since the seventies ecological studies influenced by population biology have often found continuous disturbances of populations rather than stable equilibrium, even in areas not affected by humans (Pickett and White 1985). Most studies find that populations of coexisting species can vary erratically, not around a steady mean, and a few studies have found large and unpredictable fluctuations (May 1976).

Does this mean the cultural model is wrong? Addressing the balance part of the cultural model, environmental historian Donald Worster (1990) argues that, even if nature is more unpredictable and turbulent than suggested by a balance of nature, modern human disturbances are nevertheless far more destructive than most natural fluctuations. Our background chapter reviewed estimates that the normal rate of natural species extinctions is less than one per year, but with human impacts, today's rate is more than four thousand species extinctions per year. In other words, it may be reasonable for the cultural model to qualitatively distinguish the high level of disturbance by humans from a relative balance in ecosystems lacking humans.

The cultural model of chain reactions also corresponds imperfectly to scientific ecology. Ecologists find that most organisms, in most ecosystems, exhibit substantial functional redundancy (O'Neill et al. 1986). One extreme example of redundancy is the thousands of species of microorganisms in the forest floor that decompose plant matter. Even if a large fraction of these species were wiped out, the remainder would quickly take their places. Since precisely the same functions would be filled, other species that depend on them would be unaffected. In measures of energy and mass flow, there would be no change. Even the dominant species in an ecosystem can be functionally redundant. For example, chestnut was the dominant tree species in large areas of the U.S. eastern woodlands. When it was wiped out by chestnut blight from 1906 to 1940, it was replaced by approximately twenty other tree species (Shugart and West 1977).

However, ecological science does provide some examples of just the types of chain reactions mentioned by our informants—although scientific ecology does not use that term. In a classic study of a Washington State tidal area, experimentally removing a single species—starfish—caused the number of barnacle, mussel, limpet, and chiton species to drop from fifteen to eight (Paine 1966). Moving a new species into an area to which it is not native can also have dramatic effects on native species. In a Hawaiian study of the invasion of a new tree species (*Myrica faya*) in volcanic areas, this single new species changed the characteristics of the entire ecosystem (Vitousek and Walker 1989).

We will call cases like soil microorganisms and chestnut *functional redundancy,* and cases like starfish and the susceptibility of the Hawaiian system prior to invasion *fragile interdependence.* If one were to count species, those with functional redundancy would far outnumber those with fragile interdependence. There are principles by which an ecologist can predict ecosystem relationships of fragile interdependence, but these predictions are sometimes wrong. Another factor involves the perspectives of different ecologists: those who look at the flows of mass and energy see ecosystems as very stable, even as species change. Those working in population biology see that such changes can cause loss of individual species and consider ecosystems as more fragile.

In short, ecologists would disagree that all species interrelationships are fragile interdependence. In fact, most are not. They would disagree with the cultural model that such relationships are so complex as to be totally unpredictable, while acknowledging that their predictions sometimes fail. On the other hand, when they look beyond their experimental study plots, most ecologists worry about humanity's cumulative effects of reducing the number and diversity of species, which in the long run reduces the amount of functional redundancy available and thus reduces the ability of an ecosystem to withstand stress or change.

So, for both fragile interdependence and unpredictability, the layperson's cultural model does not quite correspond to the ecologists'. The question that must be asked is, are these cultural models more like the Tukano case, managing effectively with what might seem like an incorrect model? Or, as has been found in some educational contexts, do incorrectly applied models simply mislead and waste effort?

In all three areas—balance of nature, fragile interdependence, and unpredictability—the American cultural model has selected one of several appropriate specific models from the scientific community. The cultural model incorporates stable equilibrium over continuous disturbance, fragile interdependency over functional redundancy, and chaotic unpredictability over predictable regularities. These three specific models fit together consistently—together, they comprise a more general multipart model of the dynamics of natural systems and limits to human control. In each of the three specific models, the alternative selected is a conservative one, as is the overall multipart model. That is, ecological science cannot consistently predict which changes will cause chain reactions and which will not. Thus one could justify the nonintervention model as a reasonable simplification because human disturbances are in fact risky, even if dire results are infrequent.

Models of the Causes of Environmental Concern

A final set of cultural models about nature that emerges from our interviews concerns the factors determining why other people care about nature. We have no questions asking about this topic on the semistructured interview, but it is raised as informants discuss other topics. The factors they cite include the devaluation of nature by modern economic and social systems, a lack of contact with nature leading to a lack of concern, and the value primitive peoples are thought to place on the environment. These models are important because they are invoked to understand the causes of differences of opinion in environmental debates and because they provide clues about people's own expectations and values regarding concern for nature.

Materialism and the Market System Devalue Nature

In the semistructured interviews, several informants complained that our society fosters excessive consumption and display of wealth to the exclusion of more important values.

> The fact that we are so materialistic proves that we haven't done everything right. Money is the god that we are going by, and that can't be the right god. Our god has got to be the environment, and the people, and the world.—Catherine (retired science teacher, environmentalist)

A related idea is that nature is valuable even though it is free, and that human economic exchange falsely devalues nature because it has no market price.

> These things that we have out there [the natural world] that God has developed for us are here for our enjoyment and, you know, it's free. That's one thing I think has happened to a lot of people in our society today. They've gone so materialistic they've passed right by what is out there that we can have for free.—James (farmer, custodian)

> Not everything does have a price. . . . Most living organisms are looked upon as extractable commodities rather than having an intrinsic right to exist. We have so devalued anything but our very narrow commoditization of so much of the planet.—Mark (legislative aide)

We did not ask specifically about these topics in the semistructured interviews. Nevertheless, since several informants raised them, we included them in the subsequent survey, using the following questions:

4 If people only think of making a profit, they won't really see the beauty that nature has to offer.

Earth First!	Sierra Club	Public	Dry cleaners	Sawmill workers
100	78	86	87	69

53 The present relationship between humans and nature is one of domination rather than partnership. We look at most living organisms as extractable commodities.

Earth First!	Sierra Club	Public	Dry cleaners	Sawmill workers
100	82	90	87	81

119 Capitalism may be the best system we know of today, but a fundamental problem with it is that it doesn't give any value to things you can't buy and sell, like the environment.

Earth First!	Sierra Club	Public	Dry cleaners	Sawmill workers
80	82	90	83	63

The survey results showed wide acceptance of the ideas that profit-seeking individuals, and our economic system more broadly, were at odds with environmental protection. As we discuss in a later chapter on policy, Americans correspondingly expect a reduction in level consumption in the future, and feel that a "less materialistic way of life" will help the environment.

Alienation and Lack of Contact with Nature

The second explanation our informants give for lack of environmental concern was infrequent contact with nature. There is a greater diversity of opinion on this belief than on the prior explanation of the economic system, and, in the semistructured interviews, it was raised by only four of our six environmentalists, not by other informants. Those raising this point assert that modern people are seldom in contact with the natural world and therefore do not appreciate it. They say that this lack of appreciation for nature leads to a lack of respect and little concern over environmental issues on the part of many people.

I think the majority of humans are completely alienated from nature. They don't have contact with nature. They spend their time indoors, and when they're outdoors, nature's an inconvenience to them. It makes them dirty, it bites them. Here everything's so cleaned up, everything's so tame.—Margaret (activist, environmentalist)

Just think of the people in cities. You think of Wall Street. . . . How much nature, how much environment do they get except going from one place to another in a car, and in a house, and then television, and bed, and again the next day, over and over. No, our country has changed. . . . at one time we all were as one with nature, but you aren't anymore.—Catherine (retired science teacher, environmentalist)

One person expresses the opposite position. Peter, a third-generation logger in rural Maine, feels that immersion in nature can make people take it for granted. While remarking that he thinks people are "more in tune with nature" than they were several years ago, he says:

I think up in this area people have a real tendency to take nature for granted. Living up here where we do, you're right in the woods basically, compared to somebody that's here from New York or Boston. They get out into the woods, which is a little different atmosphere for them, and they seem to really appreciate it.—Peter (logging contractor)

Elaborating this model further, Margaret proposes different levels of environmental awareness, citing examples from her experiences with residents of Pennsylvania and New Jersey. People can have an aesthetic appreciation for nature without being environmentally conscious, she asserts. They can be very concerned with the number of old-growth trees on their own properties, yet be completely unconcerned that a county garbage incinerator several hundred miles away is polluting the air they breathe. Some people, Margaret argues, are appreciative of the natural world but ignorant of the intricacies of its workings. Margaret takes her four children on extended trips away from their home near Philadelphia to land she owns in the woods of western Pennsylvania. Her purpose is to foster in them what she calls a "Zen awareness" that she believes comes from living in an undeveloped area.

In part because we did not ask about it in the semistructured interviews, only environmentalists volunteered that outdoor contact increased environmentalism. However, when we asked about such matters explicitly in the survey, majorities of most groups agreed.

3 If you don't appreciate the beauty of nature, then you may not be as environmentally concerned.

Earth First!	Sierra Club	Public	Dry cleaners	Sawmill workers
83	59	73	70	44

79 The majority of people are completely cut off from nature. They spend their time indoors, and when they're outdoors, nature is just an inconvenience to them.

Earth First!	Sierra Club	Public	Dry cleaners	Sawmill workers
97	56	57	47	56

Is this cultural model consistent with sociological findings? Only a few studies have investigated the relationship of outdoor contact with environmental sentiment, some finding a statistically significant but weak relationship (for example, Langenau et al. 1984). Anecdotally, several key figures in the American conservation movement, such as John Muir and Aldo Leopold, gave compelling but retrospective accounts of specific outdoor experiences that converted them to conservation advocacy. However, neither the studies nor the conservationist's autobiographies distinguish which came first—the environmentalist leanings or the outdoor experience.

Societies with Minimal Environmental Impact

So far, we have described the beliefs that the capitalist system devalues the environment and that most members of this society are alienated from nature. These beliefs have led some of our interviewees to make comparisons of contemporary American society with societies that are believed to have less impact on the environment. A few informants cite less-developed societies as positive models of how to treat the environment. Their examples are drawn from earlier periods in American history as well as tribal societies. This harking back to former times is not the "good old days" nostalgia of the aged, as the majority of informants who specifically mention a need for bygone values are under fifty. They do, nevertheless, often idealize those bygone times.

If we lived like we did a hundred years ago, we wouldn't have nearly the pollution now that we do. . . . [go back to] the old values . . . When I was a kid, you didn't have a wastebasket full of garbage in a month. Now you have one every day.—Bert (resort proprietor, hunting guide)

We could just shut down everything and start going back to . . . colonial times where everything was . . . a much slower pace. . . . We think we're getting so much knowledge, and we're advancing so much, but yet we really are not in control of it. We're actually doing more damage to things that probably will never be able to be [repaired.] . . . In a simpler society . . . you're not stressed out, you're not in a rat race. You take care of yourself, but also you take care of your friends and neighbors and [have] more community.—Doug (pharmaceutical scientist)

A few informants, including three of the environmentalists, brought up Native American cultures or other small-scale societies as a model of societies with minimal environmental impact. Marge cites a contemporary indigenous group in the Philippines who have "lived for thousands of years totally in balance with their environment," and Margaret cites "the Indian model of not making decisions without considering the seventh generation [to come]".[6] Abby also refers to "native peoples."

I think there was a certain balance in the very beginning. . . . If we want to try to get back to that, we should study the way native peoples have lived. There are some native people where we have the histories of them, and we can study and see how they did it and all. Almost always when we study native people who didn't leave big scars where they lived, who you could say were in balance, you almost always see that they had this great understanding, sort of an intuitive understanding of what they did. They never took too much, and they never wiped out a certain animal that they liked to eat because they always wanted some more there. . . . And they weren't worrying about the bottom line and making money or, you know, exploiting that to its nth degree because they were concerned about their future. . . . I wonder whether we've become spiritually depleted . . . as a society because we don't seem to think about the future.—Abby (shop owner, environmentalist)

These quotes combine folk history and folk anthropology with social criticism.

Our survey shows that the view of small-scale societies as environmentally sensitive is widely shared. About three-quarters of the public accept it.

118 Before Columbus came to this continent, the Indians were completely in balance with their environment. They depended on it, respected it, and didn't alter it.

Earth First!	Sierra Club	Public	Dry cleaners	Sawmill workers
58	78	77	80	69

One way to interpret this belief is that it is a statement about which direction American society should be moving. In this case, it would be related to the high acceptance noted earlier of a return to "traditional values and a less materialistic way of life." Nevertheless, this does not mean that Americans want to literally return to an earlier era, as shown by lack of agreement with the following statement:

15 We're advancing so fast and are so out of control that we should just shut down and go back to the way it was in colonial times.

Earth First!	Sierra Club	Public	Dry cleaners	Sawmill workers
42	11	14	10	22

The symbolic appeal of the environmentally sensitive "primitive" is understandable. Such peoples are seen as close to nature, unspoiled and uncorrupted, much like wilderness itself. A specific example of the rapid acceptance of a myth about the environmentally sensitive Native American is the case of the Saquamish man commonly known as Chief Seattle. A speech on environmental sensibilities has been widely attributed to him. While he did give a moving speech in 1854 about the intrusion of whites into Native American life, the environmental remarks widely attributed to him were in fact written in 1971 by a screenwriter, Ted Perry, and attributed to Chief Seattle in a film production (Egan 1992).

Nevertheless, the myth has spread quickly and widely. "It's a classic case of a lie going 20 miles an hour when truth is just putting on its boots" (historian Davis Buerge, cited in Egan 1992). Anthropologists might interpret the rapid spread of this myth as indicating what people want to believe and what they seek confirmation of.[7]

As anthropologists, we would agree that many societies with simple technology and low population density are in long-term balance with their environments. Nevertheless, some traditional societies have caused serious environmental damage, often undermining their own resource base. For example, Polynesians in New Zealand hunted the moas (a large flightless bird) to extinction over a 500-year span (Anderson 1990). Easter Island was deforested by its inhabitants over an 1100-year period, driving several tree species to extinction and reducing the carrying capacity of the island below the needs of its human population. The result was chronic warfare, cannibalism, and massive social breakdown (Diamond 1986).[8] Even societies that appear to be sustainable may only be using up their ecological support at a very slow pace (Edgerton 1992). Further examples can be found in Crosby (1986, 15) and Burch (1971). Many of these examples are somewhat larger scale hierarchical societies, or involve destruction occurring after human populations moved to new islands or new continents, or are a reaction to changes wrought by contact with Western societies.

In many other cases, traditional societies do have social structures and beliefs that support sustainable resource use (Nietschmann 1984; Rappaport 1968; Guha 1990), even though these beliefs are expressed in terms of kinship or deities rather than ecology. One explanation of small-scale societies as environmentally sensitive is a social-evolutionary one—the cultures encouraging sustainable resource use are those that have survived intact, and are thus the ones we see today, even if they did not consciously create their proenvironmental practices, beliefs, and social structures (West and Brechin 1991, 90).

In short, the lesson we take from small-scale societies is different from that of our informants, although perhaps both views would lead to the same recommendations. Our informants tend to see earlier societies as environmentally good and ours as bad. We would say instead that some societies have established long-term sustainable use of their environments while others have not. Both the empirical evidence and simple logic tell us that the societies that have not yet done so—including our own—must eventually either change their uses of the environment or destroy themselves.

Origins of Cultural Models of Nature

We can only speculate about the origins of the cultural models of nature described in this chapter. Many trace the historical origins of the broader trend of environmentalism to the conservation movement in the mid-nineteenth century and writers such as Thoreau, Audubon, Marsh, and Muir (Paehlke 1989). We suspect that the more specific models we document here have become widespread among the public more recently, because they seem to be at odds with this society's predominant literary and religious traditions (e.g., White 1967). One view is that these cultural models ultimately derive from scientific studies of biology, although they take on forms different from the scientific models (Oates 1989, 5, 31). Since few laypeople read scientific studies, there must be more immediate channels. Paehlke sees the writings of Carson and others in the sixties and seventies as bringing some of these ideas to a broad public. Even broader channels include public education, media reports, discussions with friends, and interpreting the stories of others.

Take for example the species interdependency model. It may be derived from school biology as well as the writings of popularly read environmentalists. Environmentalist writings on natural interdependencies date back at least to John Muir's (1911) observation, "When we try to pick out anything by itself, we find it hitched to everything else in the universe."

More recent examples range from the writing of Rachel Carson (1962) to the newsletters of today's environmental groups. The promulgation of these cultural models might be promoted by environmental advocacy organizations, whose agenda they support (Buttel and Taylor 1992, 221). Also, in environmental coverage in the news media, we note from casual personal observation that a common theme is that one human-caused change has other, unexpected consequences for other species.

One way in which these models are surely used is to make sense of reports and stories from others. These models are in turn passed to children, who now seem to be exposed to them at an early age. For example, consider a popular author for children from preschool up, Theodor Seuss Geisel, who wrote under the pen name Dr. Seuss. One of his stories concerns a mythical figure, the Lorax, who tries to stop a factory from cutting down trees and polluting. Part of this story's message illustrates interdependencies in nature, like those that emerged in our interviews. For example, one of several interdependencies in the short story was that brown Bar-ba-loots (bearlike creatures) depend on the Truffula trees.

He snapped, "I'm the Lorax who speaks for the trees
which you seem to be chopping as fast as you please.
But I'm *also* in charge of the Brown Bar-ba-loots
who played in the shade in their Bar-ba-loot suits
and happily lived, eating Truffula Fruits.

"NOW . . . thanks to your hacking my trees to the ground,
there's not enough Truffula Fruit to go 'round.

. . .

"They loved living here. But I can't let them stay.
They'll have to find food. And I hope that they may.
Good luck, boys," he cried. And he sent them away.

(Geisel 1971)

Understanding this story requires a cultural model of interdependency, and the story may help the reader to develop such a model if it is not already present. In recent years, environmental themes have appeared frequently in children's and adolescents' stories. The Dr. Seuss example shows that such stories have been in circulation over twenty years, and that even preschool children are expected to possess (or be able to construct) the cultural model of species interdependency.

We cannot sort out the relative timing or relative import of sources such as the popular writing of scientists, schooling, news reports, environmental advocacy organizations, personal conversations, interpretation of environmental stories, and others. Nevertheless, the current pervasiveness of stories that require these models for comprehension demonstrates—as do our surveys—that the cultural models of nature discussed in this chapter are widespread and thoroughly integrated into American culture.

Conclusion

This chapter has presented our findings that Americans possess general models of interrelationships in nature and humans' relation to nature. These models make possible elaborate inferences about environmental issues. We have documented at several points the way in which these cultural models selectively pick from scientific findings, sometimes ignoring those scientific models that would be contradictory. In the case of scientific ecology, the cultural models selected tend to be conservative, that is, the selected models provide a folk-theoretical rationale for opposing large human changes of the environment.

The correspondence of American cultural models with the findings of biology and social science is less important than the function of the cultural models in their social context. The opening of this chapter noted that we did not initially anticipate or explicitly elicit these models. Informants appealed to them in order to answer our fundamental questions, such as why they thought that protecting the environment was important, or how they could justify their environmental values. The findings described in this chapter, we will show as the book progresses, are nothing less than this culture's conceptual basis for environmentalism.

4

Cultural Models of Weather and the Atmosphere

This chapter covers cultural models that laypeople apply to the specific problem of global warming, a problem that also connects with cultural models for ozone depletion and the weather. For example, this chapter will help explain why many people incorrectly believe that global warming is caused by aerosol spray cans, that pollution controls are a good way to combat it, and that the warm weather signaling global climate change is already observable. Our discussion of cultural models of weather will explain why a potentially implausible hypothesis—that we small humans are changing the entire planet's climate—has been so quickly given credence. As in the previous chapter, we will occasionally contrast lay cultural models with those of scientists, although we will see again that lay cultural models incorporate fragments of scientific models.

Science reporting on global warming and related problems has been reinterpreted to fit existing cultural models. In part this is because global warming is relatively new, and using existing models to understand a new problem is often a reasonable cognitive strategy. As a result, however, the models discussed in this chapter seem to lead people astray more often than do the general models of the environment discussed previously.

Global Warming Incorporated into Existing Concepts

Were cultural models used to understand global warming, as we saw them being used in the previous chapter? Although global warming had been covered extensively in the news prior to our semistructured interviews, we could not assume that our informants had heard of it. Thus, after the general environmental questions, the semistructured interviews asked "*Have you heard of the greenhouse effect?*" We chose this term over global warming or global climate change as it is most commonly used by the media and seemed to be more widely recognized by the public. Seventeen of our twenty lay informants had heard of the greenhouse effect.[1] This is consistent with the proportion found shortly before our interviews by a national probability sample, 79 percent of whom had heard of the term (RSM 1989).

For those who had heard of the greenhouse effect, we asked "*What have you heard about that?*" The responses did not make sense in terms of the scientific models of global warming. To understand the responses, we hypothesize four previously existing concepts used to interpret global warming: pollution, ozone depletion, plant photosynthesis and respiration, and seasonal and geographic temperature variation. All these preexisting concepts affect current interpretation of global environmental change.

Greenhouse Gases as Pollution

From the semistructured interviews, we infer a cultural model for pollution. This model is applied to global warming, as well as to other environmental problems. The following quotation illustrates the interview data from which the pollution model is inferred.

> Now they're polluting the air by burning the garbage. And they say that they have equipment that prevents the air from being polluted as a result of the garbage being burned. On the other hand, the people say that their cars are losing their paint in the immediate area of the incinerator and that the paint is peeling off the houses, and I suspect that in due course, several years down the road they will discover that the incidence of cancer around the incinerator, which fortunately is located some distance from us here, is on the upswing.—Ronald (resort proprietor)

From statements such as this from many different informants, we infer a cultural model of pollution that includes elements such as the following:

1. Pollution consists of artificial chemicals, not natural substances.

2. These chemicals are toxic to humans and other life, although health effects may not be observed until much later.

3. Sources of these artificial chemicals are predominately industrial and automotive.
4. Pollution is fixed by installing additional filtering equipment.

Greenhouse gas emissions are conceived as just another instance of the general pollution model. However, applying the pollution model to climate change leads to incorrect inferences about, for example, health effects.

Well, I like warm weather, personally, but I think it's wrong for what humans are doing to the atmosphere. *In what way?* With all the aerosols and the ozone and so forth . . . that's being projected up into the atmosphere. . . . [*If you like warm weather, why do you say the greenhouse effect would be wrong?*] Well, I think it's wrong because at the same time, we are ingesting and breathing in all these different chemicals that are being put into the atmosphere.—Susan (hospital administrator)

A chemist would note that the pollution model is leading this person's reasoning astray because the major gases causing both global warming and ozone depletion (CO_2 and CFCs) are nontoxic. Some lesser greenhouse gases (e.g., methane) are toxic but are far too dilute in the atmosphere to have any effect.

Another way in which the general pollution model misleads when applied to global warming and ozone depletion is by concluding that traditional pollution controls seem to be a solution.

[To prevent global warming] we're just going to have to probably . . . find out where most of the pollution is coming from. [For the sources that are] industrial, [we will need to] have an incredibly fine filter . . . where you prevent most of this excess CO_2 from going into the atmosphere.—Doug (pharmaceutical scientist)

Filters work for some types of particulate air pollution, however, greenhouse gases are not particles. No control technology exists for CO_2. Unless and until some such technology is developed, the concept of emissions control is inapplicable. In fact, current pollution-control technologies most often *increase* CO_2 emissions. For example, scrubbers on coal power plants reduce traditional air pollutants such as sulfur dioxide, but in so doing they make the plants slightly less efficient. Thus, more coal must be burned to compensate, producing more CO_2. Most analysts believe that CO_2 reductions will have to be achieved through energy efficiency and switching to nonfossil fuels, not by use of filters or other pollution controls.

Another problem with the pollution model is that it focuses on industrial smokestacks and vehicle sources. Although burning of fossil fuels contributes 55 percent of anthropogenic global warming, fitting global warming into the cultural model of pollution obscures the roles played by invisible and seemingly nonpolluting sources such as farming, land clearing, and end-use energy inefficiency.

Of the informants who provided enough detail to discern, about two-thirds applied pollution concepts inappropriately to global warming. Consequently, the pollution model is a source of many misconceptions about global warming, which also leads to ineffective policy choices.

Ozone Depletion

The most widely shared transformation from scientific to lay knowledge is caused by categorizing global warming as a subset or a result of ozone depletion, as in the following example:

> *Have you heard about the greenhouse effect?* Is that what they're talking about the ozone layer? . . . [That] last year . . . create[d] the hot spell? . . . But I couldn't understand that, last year we had it, what made it change this year? We don't have it quite as severe. . . . *What other things can you remember?* . . . Well that was about the only thing . . . through the gases and that in the cans, you know, pressurized cans.—Wilbur (retired fireman)

Although the man quoted above had only a ninth-grade education, the same confusion is seen in the following quotation from a college graduate, who expresses an interest in environmental and health news and who warned us that she was atypical because she talks with scientists about the greenhouse effect.

> Most people say burning fossil fuels is changing the climate because we are making the ozone layer disappear—that's the layer that protects us from the sun's harmful rays. This will greatly affect the climate over the next hundred years.—Ellen (freelance writer)

There are some interdependencies among climate change, tropospheric (near surface) ozone pollution, and stratospheric ozone depletion. For example, CFCs cause stratospheric ozone depletion as well as global warming, and ozone itself is a greenhouse gas. However, these are secondary, tertiary, or lesser effects: the lay model is clearly different from the scientists' model when people give the ozone layer or pressurized cans as the primary, or the only, cause of global warming.

One informant constructed a model that explicitly connected the two.

Have you heard . . . what's causing the greenhouse effect? . . . the amount of protective atmosphere around the world . . . the layers are getting thinner and thinner, and as more heat gets through that tunnel . . .—Peter (logging contractor)

We believe that informants' confounding of ozone depletion and the greenhouse effect, like that in Peter's statement, is due chiefly to the overlapping of features of these two human-caused atmospheric global changes and the simultaneous presentation of the two in news reports. Further, we believe that information about the greenhouse effect is assimilated to a model of ozone depletion because the ozone model was established first. We regard Peter's description of the ozone hole letting heat through as a post hoc explanation that allows him to assimilate the two phenomena. Emma's introspections also show some of these features.

What have you heard about the greenhouse effect? Well, I was confused about it for a long time, cause I had heard about the CFC part first. . . . I think, in this sense, I am a lot like the average lay person. You know, I might see some broadcast here, and I might read some article there, and something talks about ozone holes, and something talks about acid rain, and something talks about . . . the greenhouse effect itself. And, at first, I was confused. [I wondered if] all of these things were unitary really . . . all contributing to the same problem? Or are these problems that are separate, but work together to make the whole atmosphere less healthy, less conducive to us? And I still am a little confused. But I think I got to the point where I thought that the CFC stuff was related to the CO_2 stuff. But if they were both part of, they were both greenhouse gases, I'm just . . . [indicates confusion] I really at this moment [pause] are you going to "X" that out on my test questionnaire?—Emma (coal loading machine operator)

Despite her embarrassment, Emma is more aware of the distinction than most we interviewed.

Occasionally, even major national media confound these two phenomena. For example, a *U.S. News & World Report* article described "ozone-depleting carbon-dioxide emissions,"[2] and the *New York Times* twice in 1992 referred to carbon dioxide as damaging the ozone layer (Nassar 1992; Lewis 1992). These confusions were made by writers who were not primarily science writers—people who, after all, are as susceptible as any of us to overextend familiar categories. However, we do not believe that the public's confusion on this issue is due to these occasional errors by the media, because the initial reports that we reviewed in major print media, prior to our interviews, did correctly distinguish between the two.

Additionally, some beliefs about ozone depletion itself are incorrect: many informants assigned the blame to aerosol cans (a notion that is a decade out of date in the United States), and some confused stratospheric ozone depletion with urban tropospheric ozone pollution.

In sum, it appears that Americans are now fairly familiar with ozone depletion, an earlier and simpler concept with fewer causes and fewer consequences than global warming. The assimilation of the greenhouse effect to a model of ozone depletion is an example of *syncretism,* a process often documented by anthropologists in which new information is assimilated to fit existing familiar concepts.[3] The American view of ozone depletion being caused by spray cans is probably a simple cultural model. However, we have refrained from calling the overlap of ozone depletion with the greenhouse effect a cultural model, as this overlap may be a simpler case of placing a new phenomenon into a familiar (if inappropriate) category. But regardless of how we describe it, this lay conceptual problem leads to inappropriate inferences.

Photosynthesis and Respiration

New information about the greenhouse effect is also incorporated into a simplified cultural model of plant photosynthesis and respiration. The cultural model draws some elements from scientists' models, but it contains fewer elements and leads to several highly divergent conclusions.

Several informants understood and could recite the idea that trees absorb carbon dioxide and produce oxygen. From this foundation, some had transformed the media's descriptions of how deforestation is raising levels of carbon dioxide into the idea that humans would exhaust all atmospheric oxygen, a very frightening prospect.

Well it has to do, I think, with the climate changing as a result of the atmosphere, atmospheric changes that are presently going, and cutting down all of our woods takes away a certain chem—the oxygen, or something, that is required for us to have good air quality, and it's kind of scary.—Tara (sales manager)

That's what scares me. *What?* When they cut all the forests down, they say, pretty soon we're not going to have any oxygen to breathe. Why do they let them do that?—Cindy (housewife)

Worries about breathing were volunteered by only a handful during the semistructured interview. Nevertheless, our survey showed that it is of concern to substantial majorities of all groups save those in the timber industry.

94 If they cut all the forests down, we would soon run out of oxygen to breathe.

Earth First!	Sierra Club	Public	Dry cleaners	Sawmill workers
64	58	77	67	44

To the best of our knowledge, no national probability sample has ever been asked about this indicator of overextending a photosynthesis model, even though our small samples suggest that it is very common. On the above question, for example, over two-thirds answer incorrectly, giving the answer consistent with improper use of a photosynthesis model.

An educator or atmospheric chemist might initially try to correct this problem by just pointing out that the quantities involved could not cause a problem. CO_2 is currently only 0.034 percent of the volume of the atmosphere, compared with O_2 at 21 percent (Harte and Socolow 1971, 281). A doubling would raise CO_2 to 0.068 percent, or an ultimate sixfold increase to 0.2 percent. O_2 would still be over 20 percent, nowhere close to a danger of suffocating humans or other mammals. Even a sixfold increase in CO_2 would be less than the CO_2 in many urban office buildings, and far less CO_2 than people breathe in a confined environment like a submarine or the space shuttle. But a pedagogical strategy of correcting only the quantities is dealing only on the surface, we feel, and will not change the underlying cultural model that generates the specific errors.[4]

We infer from our interviews what we believe to be the underlying cause of the concern for running out of oxygen: people are using a simplified model of photosynthesis and respiration, in which the oxygen we breathe is thought to come directly from today's living plants.

Figure 4.1 illustrates the cultural model we infer people to be using. People inhale O_2 and exhale CO_2; plants take in CO_2 and give off O_2. From this model, the conclusion "if they cut all the forests down, we would soon run out of oxygen" follows directly. We find figures precisely like this in secondary school biology textbooks illustrating the oxygen cycle (for example, Smallwood and Green 1977, 642).

There are three problems with this model, beginning with an omission immediately visible in the figure: the figure shows carbon atoms flowing only one way but O_2 atoms circulating both ways (from plants as O_2 and to plants as part of the CO_2). The two derivative problems are that the figure does not show what happens when the tree dies, or how humans get the carbon in the CO_2 they exhale.

A more complete model of how scientists understand the flows would use a food crop as the illustrative plant and show the circular flow of carbon, as in figure 4.2. This figure makes clearer that part of the carbon that plants store from CO_2 is being recycled as plant food that we eat. A more detailed figure could show that the plant is using water and fertilizer, that the carbon in the corncob is not just "C" but carbohydrates—compounds of CH, that the person is also exhaling water vapor and excreting carbon, which is then consumed by decomposers, and so

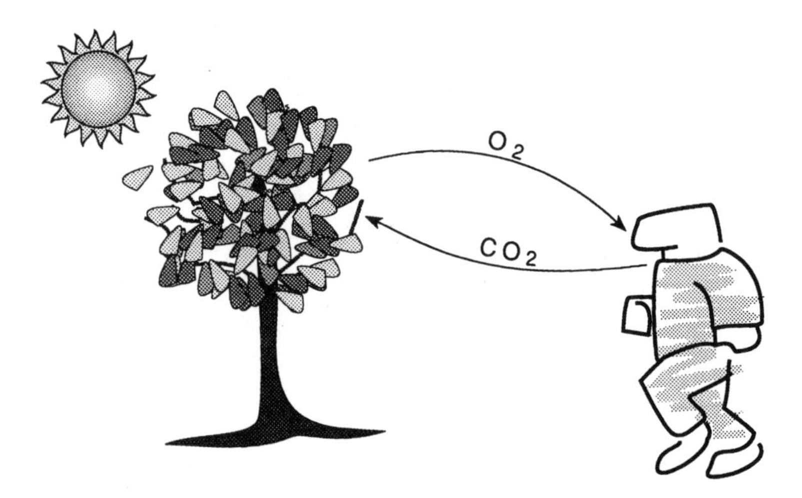

Figure 4.1
Inferred cultural model of photosynthesis and respiration

Figure 4.2
An augmented example of photosynthesis and respiration, showing the return path of carbon

forth. More detailed figures, showing the return path of carbon in food and other flows, can be found in science textbook descriptions of the carbon cycle or the carbon-oxygen cycle (See Mallinson et al. 1991: 108). Also, both figures here are simplifications in that the O_2 we breathe is actually a mixture from many different plants producing O_2 over millennia.

Even the revised diagram is misleading in that it does not include the entire life cycle of the plant. During growth, trees and other plants do take in CO_2 and give off O_2, as shown in figure 4.1. Plants accumulate carbon as they grow. When they die and decompose, all that carbon is given off—whether sooner, as the tree decomposes in the forest, or later, as buildings built from timber eventually rot or burn. (Rotting in fact means small organisms using the dead tree for food and giving off CO_2; burning is approximately the same thing occurring more quickly through a chemical reaction.) Thus, a fully mature standing forest, with its rate of growth equal to its rate of decomposition, should theoretically add no O_2 to and subtract no CO_2 from the atmosphere. The growth of young trees absorbing CO_2 is balanced by decomposition of dead ones giving it off again.

Felling a standing forest to burn or rot will release a one-time shot of its stored carbon, but the carbon released by deforestation is small in comparison to fossil fuels. For comparison, burning all living matter on the planet would release 560 gigatons (GT) of carbon (Clark 1982, 433), not even enough to double atmospheric CO_2. Economically recoverable fossil fuels will multiply atmospheric CO_2 by six times. Total deforestation now accounts for only 9 percent of the forcing of climate change. From this perspective, it would seem that an environmentally concerned person would be concerned about deforestation not so much for its climatic effects, but because it is an indicator of how quickly humanity is destroying the habitats that support a diversity of species.

Only over geological time scales can we say that plants (including trees) have significantly increased O_2 and reduced CO_2 in the atmosphere. Plants that are buried or otherwise isolated, and therefore do not decompose, remove carbon from the atmosphere-biosphere system.[5] This is how fossil fuels are formed; humans reverse this process by burning fossil fuels. Because burying without decomposition occurs rarely in nature, it accounts for only 0.04 percent of the O_2 produced by plants each year, or one part in 15,000,000 of atmospheric oxygen (Broecker 1970, 1531). Plants have been at this slow task of producing the atmosphere's inventory of O_2 from CO_2 for over two billion years.

In short, it is correct that ultimately plant photosynthesis and respiration, in conjunction with landslides, sedimentation, tar pits, and other natural occurrences have produced the atmosphere's O_2. Humans reverse this accomplishment to a greater degree by burning fossil fuels than by burning forests. But neither activity could cause us to suffocate. We could not destroy enough plants to deplete our O_2 supply because there is too much atmospheric O_2, and, as suggested by figure 4.2, because we would starve first. And we could not burn enough fossil fuel to suffocate because the economically recoverable fossil fuels are small in comparison to the atmosphere's O_2 (much more fossilized carbon exists in low grade, uneconomical deposits).

The layperson's model of photosynthesis is an especially good example of what we mean by a cultural model—we infer the simplified photosynthesis and respiration model from semistructured interviews because it explains many different specific beliefs divergent from scientific models. Our survey suggests that this model is held by a substantial majority of the public. We call it a model because it includes multiple components and rules about how components interact and can produce many separate individual predictions. As with other American cultural models of the environment, it draws from popularizations of science rather than being constructed independently like the Tukano example we began with. The photosynthesis and respiration model is taught early in school and is widely shared. It may be reinforced by mass media, in conversations with friends, and through other sources.

Two practical results flow from Americans' use of the oversimplified model of photosynthesis and respiration. The first is undue concern for running out of oxygen because of deforestation. Second, this model may contribute to an exaggerated attribution of deforestation (vs. fossil fuels) as a cause of climate change, reversing the true ranking presented in our background chapter. This analysis suggests to us some room for improvement in environmental education such as the teaching of biology, museum exhibits on global warming, and environmental television programs.

Significance of Temperature Variation

The fourth source of the deviation of cultural models from scientific ones is preexistent lay knowledge of seasonal and geographical temperature variation. North Americans are familiar with winter-summer temperature swings of 100°F, frequent daily temperature swings of 20°F, and major geographical differences in temperature. From this experience, an average long-term temperature rise of less than 10°F does not seem very harmful. This is seen in some of the reactions to the briefing on global warming (near the end of the semistructured interviews).

That doesn't sound so bad, does it? Three to nine degrees [warming in a hundred years].—Cindy (housewife)

I certainly don't care as an individual whether its nine degrees warmer. Living in New Jersey it can be forty degrees different than it was yesterday.—Nick (small manufacturing plant owner)

One person argued that future generations would acclimatize to the new temperatures.

I wouldn't want to live in that, from what I am accustomed to. But it relates, I'm sure, to what you're accustomed to. If you live in Alaska, or you live at the North Pole, and you like it, [then] you like it.—Tara (sales manager)

Other support for our observation that people underestimate the effects of seemingly small temperature changes derives from anecdotal evidence provided by a climatologist. Schneider (1988) notes, "Many people are surprised to find out that the ice age was only about 9°F colder than the present average earth temperature." Although he did not say so in the article, Schneider based this observation on questions asked by audiences at his public lectures and testimony, a subpopulation presumably more informed than the average citizen (Schneider, personal communication).

In other words, some people find it difficult to understand what all the fuss is about because their direct experience, from both travel and local temperature cycles, teaches that 3°F to 9°F is not much temperature change. However, our survey suggests that this misconception leads only a minority to feel sanguine about global temperature shifts.

113 There is no reason to be concerned about the effects of a five-degree increase in average temperature because daily temperature varies by more than that from night to day.

Earth First!	Sierra Club	Public	Dry cleaners	Sawmill workers
0	7	13	13	26

A related difference between scientific and lay knowledge is the understanding of the significance of higher average temperatures. When a scientist, especially a climatologist or biologist, refers to concern about global warming, he or she implicitly is referring to multiple geophysical and ecosystem effects. These include changes in overall weather patterns, more violent storms, rise in sea level, possible shift in air and water currents with concomitant major regional climate changes, shifts in ecological zones, and change in the lands suitable for farming. To laypeople, global warming without further elaboration simply means hotter weather. When our semistructured interviews first asked about global warming, warmer weather was the first mentioned and, for several, the only consequence the informants related to the greenhouse effect.

What have you heard about [*the greenhouse effect*]*?* This portion of the country basically is going to become warmer over a certain length of time. That's really all I know about it. . . . *Have you heard about any other effects . . .?* Not that I can recall.—Susan (hospital administrator)

While middle-class informants such as Susan did not see warmer weather as a great problem, it was more troublesome to Paige, who draws on his experiences with life in a lower-income neighborhood of the inner city.

If the weather gets a lot warmer, do you think it would be good, bad, or neutral? I think it would be bad; I think it would be terrible. *Why?* Well, I think people react differently in warm weather than when it's cooler. I think it has an effect on attitudes—behavior. . . . I mean in the prison system especially, where the people are just, you know, stuck in there, and they've got to let off steam. So, sure. *So you think in prison it makes people more violent?* Sure, but outside the prisons, too, 'cause I even see it at work; you know, when the weather is extremely warm, people tend to be, you know, a little hot tempered. I think, you know, their blood boils. And when the blood boils in the body, it goes to the head, and next thing you know, there's, you know, an explosion. . . . I've seen them react that way.—Paige (manufacturing worker)

Without endorsing all the explanatory mechanisms advanced by Paige, it is clear that some people perceive hot weather as affecting mood and hence having personal and social costs beyond discomfort. Nevertheless, Paige resembles other informants in conceiving of global warming in terms of the personally familiar effects of hot summer weather rather than the more profound, but unfamiliar, environmental consequences.

In sum, many people already have a sense of how much temperature fluctuates and what effects hot weather has upon humans. This mixture of simple concepts and cultural models about temperature is used to interpret global warming.

In contrast to laypeople, scientists do not generally consider human response to hotter weather to be a major problem of global warming.[6] The primary concerns are the indirect effects of temperatures: shifted biological range, changes in agriculture, increased frequency and severity of tropical storms, higher sea level, and so on. What seems to be missing from the lay understanding is that changes in *average* temperature can have huge effects on nature, even if the average changes by much less than the daily or seasonal *range* of temperatures. What is important is not warming per se but climate change and the resulting environmental change. The confusion is augmented by use of the term *global warming* for climate change, which focuses attention on warmer weather, probably the least significant of all the potential climatological, storm, and ecosystem effects.[7]

Effects and Durability of Existing Concepts

Our exposition of lay concepts and models, we believe, explains many specific differences between laypeople and scientists regarding individual facts. These discrepancies have not yet been tested on national probability samples, although one local survey in Pittsburgh confirms some of our findings. Using an open-ended question that allowed multiple responses, Read et al. (1994) asked for "things that could cause global warming." The responses are shown in table 4.1. Various synonyms for deforestation, sources of pollution, and CFC-induced ozone depletion head the list. Despite their sample being highly educated (62 percent college graduates), these data provide independent confirming evidence for the preexisting concepts and cultural models we propose in this chapter.

In summary, the application of these four preexisting concepts to global warming illustrates the difficulties in introducing a new idea into culture. An initial publication discussing the distorting effects of these four concepts (Kempton 1991a) speculated that continued media coverage, school lessons on this topic, and public discussion would improve American citizens' understanding. Subsequent analysis of prior environmental misconceptions (Kempton 1993) makes us less sanguine. For example, although Americans had assimilated the (former) link between spray cans and ozone depletion by the late seventies, few have connected ozone depletion with air conditioning and refrigeration, even given a subsequent fifteen years to do so. This parallel example is all the more worrisome because the causes of ozone depletion are simpler than causes of global warming. Until existing cultural models are augmented or

Table 4.1
Responses to an open-ended question about "things that could cause global warming" from Read et al. (1994)

Cause of global warming	Percentage mentioning
Reduction of biomass	57
Automobiles	41
Industry	32
Pollution	30
Depletion of ozone layer	27
Aerosol cans	26
CFC emissions	20
Burning fossil fuels	18
Gases/chemicals	18
Nuclear power/weapons	15
CO_2	14
Natural causes	14
Overpopulation	11

refined, the scientific content of media discussions, no matter how accurate the words and graphs employed, will continue to be distorted as citizens try to fit this strange new phenomenon into their known world. Some of this refinement of cultural models will probably occur on its own, but other gaps and problems may be very persistent without deliberate, targeted educational efforts.

Perceptions of Weather

People directly perceive immediate weather conditions, not the long-term climatic regime. Some climatologists are concerned that citizens' inability to directly observe climate, and their faulty inferences about climate, would limit public support for action on climate policies (Schneider 1990). There are three perceptual or cognitive questions that precede the question of whether citizens will apply political pressure: (1) how do people generalize from multiple, sporadic, casual weather observations to draw conclusions from them; (2) if the climate begins to change, would the average citizen notice; and (3) would ordinary people find plausible the claim that climatic warming was caused by anthropogenic gases?

To address these questions, we briefly explored informants' personal observations about weather patterns. The semistructured interviews began by asking about weather so that informants' answers would not be influenced by our later questions on environmental pollution and global warming. Our data suggest that people overemphasize both the human effects on climate and the extent to which the climate has already changed.

Human Effects on Weather

As a quick check on the informant's knowledge of meteorology, we began our interview by asking: *What factors would you say affect the weather?* We got some answers consistent with meteorology.

> The jet stream is the biggest factor concerning the weather . . .—Jenny (social studies teacher)
>
> Sunspots, volcanic activity, earth movements.—Ellen (freelance writer)

A more common type of answer was one we did not anticipate. People attributed changes in weather patterns to diverse human perturbations. Several mentioned pollution as the primary factor.

> Pollution affects the weather. That's all I can think of.—Paige (manufacturing worker)
>
> Well, burning, like these rain forests, and these Western forest fires. Spraying for insecticides and stuff like that. And, herbicides, like on the farms, to prevent weeds from growin'. . . . But the main thing is burning and auto pollution, stuff like that.—Walt (retired machinist)

The greater number of responses describing human effects on weather as opposed to natural ones may have been exaggerated by our question wording, but the frequency with which human effects were mentioned certainly illustrates that people consider human effects on the weather to be plausible.[8]

One man thought the weather had become more violent or erratic, a change he attributed to atomic bomb testing.

> I have an answer, but it's, [pause] I don't know, I've always felt that when they had that bomb I think it had an awful bearing on the change of our weather.

The A-bomb. They had those tests. . . . just seemed like here things have changed ever since, it's become more torrent, the weather here in the past few years. . . . *When you say "more torrent," you mean like more changing?* Violent, violent, yeah. The weather is very changeable. They say that didn't have nothing to do with it, but I still feel that it did somewhere along the line.—Wilbur (retired fireman)

Lay attribution of weather change to atomic bomb testing was in fact widespread in the early sixties (Kimble 1962). A national news magazine's article at the time began: "Many remain convinced that man—especially American man—has knocked the weather out of kilter with space and atomic experiments. Officials say no." The same article recalled that during World War I, unusually wet weather in Europe and the United States was widely attributed to extensive artillery fire (*U.S. News & World Report* 1963, 46, 48).

In our semistructured interviews, three people (15 percent of our lay sample) mentioned space shots as affecting the weather, two of whom are quoted below.

I have my own private theory. [pause] *What's that?* That every time they shoot something up in space it disturbs things up there! *There could be something to that.*[9] I've been told I have no foundation for that, but it just seems every time something happens we get this strange type of weather. . . . *Like what?* . . . Well, for instance, tornadoes were very rare in this section of the country . . . tornadoes, and violent-type storms. . . . It used to be rather calm here.—Susan (hospital administrator)

(In response to a request for ideas as to what could be done about the greenhouse effect:) Well, I don't know what the hell they're doin' up on the moon and shooting those things up there. I think they're disturbing the atmosphere. So much rain we've had, so much rain.—John (retired factory foreman)

The survey illustrates that a higher percentage admits the possibility of human effects on weather when we raise the possibility.

97 There may be a link between the changes in the weather and all the rockets they have fired into outer space.

Earth First!	Sierra Club	Public	Dry cleaners	Sawmill workers
79	33	43	31	48

The quotations and survey data both suggest that people are predisposed to believe that the weather is affected by human activities, especially human activities that occur in the atmosphere and are regarded as unnatural or immoral (space shots, atomic bombs, extensive artillery fire, pollution). Thus, it is not surprising that informants seemed to readily accept that human emissions of greenhouse gases could cause climate change. We do not completely understand the beliefs or models underlying this, but note that something culturally important is operating when such beliefs persist in the face of denials by authorities. The persistence is seen in *U.S. News*, "Many remain convinced . . . officials say no," and Wilbur, "They say that don't have nothing to do with it, but I still feel it did . . . ," and Susan, "I've been told I have no foundation for that, but it just seems . . .". In other words, there seems little basis for the concern of some who speak about the greenhouse effect (Schneider 1990) that laypeople would find human-caused weather changes implausible—if anything, they are more ready to accept human influence than are most scientists.

We also see here hints that these beliefs are not value neutral. For some, it is clearly improper for humans to change the weather. This is explicit in John's quote ("I don't know what the hell they're doing . . .") and is suggested by others. More direct evidence can be found in earlier surveys by Farhar. In nonurban areas of four states, she asked whether respondents believed that "cloud seeding probably violated God's plans for man and the weather." Agreement ranged from 30 to 48 percent (Farhar 1977, 289). Agreement with this statement also correlated with two secular statements: "Man should take the weather as it comes and not try to alter it to suit his needs or wishes" and "cloud seeding programs are very likely to upset the balance of nature" (Farhar 1976).

Would similar concerns about "God's plans" or the "balance of nature" be cited in reaction to nonweather environmental changes, say, polluting a river or paving a meadow for a shopping center? While they might have other objections to these environmental changes, we believe that few Americans would express their objections to modifications of other aspects of the environment in the same terms. It appears that weather, and hence climate, is given special cultural importance, and, even in our technology-based society, many people feel that humans should not tamper with it.

Belief that the Weather Has Changed

We find that a majority believes that climate has *already* changed. Some reported their personal observations of warming, most typically noting milder winters.

We used to have snowdrifts all the way up the telephone poles, all the way up the side of the house. Now, you get a couple of inches and they close the schools. *When do you think that changed . . .?* I guess in the last ten years, there was a big change. [*Are you thinking about this now for the first time?*] No, I've thought about it, we've discussed that. [Her mother indicates agreement.] You know when we look at old pictures and stuff we say, "Look at all that snow."—Cindy (housewife in New Jersey)

I've noticed an awful lot of cooler weather [in the summer] . . . Also I've noticed in the wintertime, our winters have changed drastically. . . . For example, fifty years ago, we'd have winters that would come in to be 35, 40, 45 below zero and last for ten or twelve days. We haven't seen that for years and years and years. . . . I think that if we don't start doing something to clear this greenhouse effect, we're gonna be in serious trouble. . . . Also the last five years here in this area have been real mild . . . we just haven't had the snow. . . . Something has happened and there again it must be the pollution we're getting in the air.—James (farmer, custodian, in Maine)

We will cite literature later in this section that compares such reports with actual weather patterns and finds the reports inconsistent with actual weather changes. The preceding comments give no evidence that they are anything but objective reports. The survey shows that, to our amazement and contrary to the concern of some climatologists, substantial majorities already claim to have personally observed the effects of global warming.

135 You can already notice the effects of global climate change on the weather around here.

Earth First!	Sierra Club	Public	Dry cleaners	Sawmill workers
93	64	83	83	56

It is effectively impossible for laypeople to accurately discern a climate trend from their own casual observations of local weather. Even working from global data, climatologists find the signal of an increase in global temperatures difficult to detect over the background of random fluctuations. Although some individual scientists believe they have detected warming, the consensus of an international scientific committee on this topic was that "the unequivocal detection of the enhanced greenhouse effect from observations is not likely for a decade or more" (IPCC 1992, 6).

Another common observation made in the semistructured interviews was that weather patterns were becoming less predictable.

Well your springs are unpredictable any more, you can't really tell. As far as, you know, especially if you tried to plant a garden and that. . . . Now you can't really go by it. What was it? Somewhere in the middle, fifteenth of April when you could pretty near predict that frost was done. And you go ahead and plant. But now . . . you could have a cold spot, you know, it can seem like anytime you turn right around, it's snowing.—Wilbur (retired fireman)

I think our weather is more unpredictable now, we've got more modern satellites and all to predict the weather and I think they're worse at predictin' it now than when they didn't have them. So, your seasons don't seem to run the same. I mean, they're less predictable for patterns. . . . Like now in the spring you get real hot weather like the summer, and then when the summer comes you got weather that's like the spring, and then in the middle of winter sometimes you get weather that's almost like summer. It isn't an even pattern. *When would you say you started noticing that . . .?* I don't know, it's come on gradual.—Walt (retired machinist)

Our survey verified this belief in the change in weather patterns and showed it to also be more widely held than we expected.

1 The weather has been more variable and unpredictable recently around here.

Earth First!	Sierra Club	Public	Dry cleaners	Sawmill workers
93	52	79	73	74

Besides warmer winters and more unpredictability, people who believed that unnatural or immoral human activity was influencing the weather often perceived an increase in storms or violent weather.

Although we found no national surveys asking this important question, a 1977 study in Illinois did and also compared the responses to long-term meteorological data. The survey asked: "Have you noticed any particular changes in the weather during the time that you've lived here?" Sixty-two percent answered yes (Farhar et al. 1979, tables B17, B18). Asked how they observed this change, respondents overwhelming cited direct personal observation of the weather (55 percent direct observation, 16 percent talking to others, 16 percent TV/radio, 9 percent newspaper/magazine, and 1 percent or fewer for any other response). The study concerned a local weather change that actually had occurred in the St. Louis area over the prior twenty-five years: increases in growing-season precipitation, thunderstorms, and hail. Actual weather change does not explain the finding, however, because only 11 percent reported the changes that actually did occur. The most commonly reported change was less rain, the opposite of the reality, perhaps due to the immediately prior summer being unusually dry. Also, answers were not significantly different in the control area where the weather change had not occurred (Farhar et al. 1979, D3). In short, people report changing weather, whether or not it occurs.

Our finding that a majority reported noticing a change in the weather is also consistent with historical data on perceived weather changes. For example, during the American colonial period there was a widespread, publicly articulated belief that the American climate had become warmer. Ludlum (1987) cites people's accounts that winters did not seem as harsh as those described by their grandparents. One early published explanation (Williamson 1771) attributed the warming to human alteration of the New World: felling trees allowed temperate marine winds to penetrate from the east more deeply into the country and bared soil to receive and store some solar heat. Another popular explanation was that "the rise of urban communities with heated buildings and smokepots was leading to a milder climate, as they claimed had occurred in Europe" (Ludlum 1987, 257). The attribution of climatic warming to deforestation, heated buildings, and smokepots rings curiously familiar in the present. The similarity of specific effects may be coincidental; our point is that people have a historical propensity to perceive weather change, whether or not it is occurring, and to attribute it to human perturbations.

How Can Lay Weather Observations Be Explained?

Climate is manifested by the long-term patterns of weather. It is not surprising that laypeople do not compute accurate statistical trend lines based on personal observation. What is surprising is that the majority claim to have personally observed weather changes during their lifetimes. One might try to explain the frequent observation of warming as accurate, because most of our informants, and most U.S. residents, live in areas that have experienced ambient temperature increases due to urban growth. This effect, referred to as *urban heat islands,* is carefully factored out by climatologists studying the weather record, but it is the strongest signal present in casual personal observations. However, such an explanation could not explain either the (rural) Illinois study nor the Ludlum colonial data.

Several psychological factors could explain these findings. The common perception that the weather is more variable now than it used to be may be due to what cognitive psychologist Eleanor Rosch has called the "good-old-days effect" (personal communication), that is, long-term memories are systematically distorted to conform to an expected typical pattern, whereas in the recall of recent experience, deviations from typical patterns are more noticeable. This would be an example of a cognitive illusion in which we remember a distant past when the summers were consistently hot and the winters cold, just as they should be. The impression shared by many of our informants that global warming has already started may also be a result of overweighting the couple of years of hot summers and mild winters in North America immediately before our interviews. (If so, the subsequent couple of years of the Pinatubo volcano-induced cooling would have cast this conclusion into doubt. In neither case would the change be broad enough in time or place for a climatologist to legitimately infer global warming.)

A further possibility is that their overestimation of human influence and control over the weather and environment may stem from another cognitive illusion described by Tversky and Kahneman (1982) in their work on judgment under uncertainty. In emphasizing the possible role that A-bomb tests, pollution, and space shots have in affecting the weather, people inflate the importance of the factors that humans have responsibility for (and hence some control over) above those factors that are neither caused nor controllable by humans (e.g., random events, the sunspot cycle).

In sum, these data do not provide any basis for the concern expressed by some climatologists (for example, Schneider 1990) that people would not notice global warming or would not attribute it to human activities. On the contrary, the evidence presented here suggests that at any given historical moment, most of the population believe that they have observed a long-term change in the weather. Warmer winters and more variable and violent weather seem to be common observations. The current publicity about global warming gives people an appropriate framework within which they can interpret their own observations. Further, many people are predisposed to attribute changes in weather to unnatural human activities. These factors, together with the credibility granted scientists (described in the chapter on policy reasoning), clarify why a potentially implausible hypothesis—that human activities will warm the entire planet—has been so readily picked up by the general public.

Conclusion

As in the previous chapter, this chapter has shown that people do not just passively receive new information, but rather actively fit it to their existing cultural models and concepts. For example, we found that people currently understand global warming by reference to prior experience with natural temperature fluctuations and to three cultural models: pollution, ozone depletion, and photosynthesis and respiration. Further models may be applied with less frequency.

One of the practical consequences of these models is that they misdirect concern. Based on our informants' reports, we would suppose that there are millions of Americans forgoing aerosol cans out of their concern for the greenhouse effect, an ineffective measure. As we will show, conclusions drawn from the use of inappropriate models also lead to support for ineffective policies.

We therefore recommend that anyone trying to communicate with the public about global environmental change address the preexisting models and concepts rather than assume they are writing on a blank slate. We suspect that this advice applies to public communication about most areas of science. Recent work on risk communication has shown that notably better results are achieved when communications are designed on the basis of cultural models research on how people understand the subject of the communication. Models-based communication performs better even when measured against communication designed by respected communication experts, working from general principles (Morgan et al. 1992a; 1992b). We will discuss this further in the book's conclusion.

Regarding weather observations, we argued that weather frequently seems abnormal to people, that there is a propensity to blame weather changes (which may not even exist) on certain types of human activity, and that people feel it is unwise, perhaps even immoral, for humans to deliberately modify the weather. These factors have undoubtedly raised the public profile of global warming, advancing the agenda of activists in this field, unbeknownst to the activists themselves. Do these weather interpretations make the public overly susceptible to calls for action on global warming? They could, or they may only help to overcome the disadvantages of this issue's complexity and invisibility.

5

Environmental Values

Our research was designed to go beyond factual beliefs to values in order to understand what would motivate environmental concern and action. This chapter explores the values Americans hold about nature and how humans should treat nature. We use the term *values* to refer to moral guidelines. *Moral,* in the discipline of ethics, refers to questions concerning the well-being of other human beings or other living cretures, not just oneself (Beauchamp 1982). Values can operate in conjunction with other motivations. They also can apply in situations where political and market solutions are ineffective. For example, someone may halt an environmentally damaging practice both because it saves money and because they feel it is the right thing to do morally. Values are especially a concern for global environmental change, because the causes of damage are disconnected in time and space from those who are harmed (i.e., people one will never meet or who do not yet exist). Thus environmental values may be the only reason to prevent global environmental change, because there are no economic motivations or pressure from harmed political constituencies.

Based on our interviews and survey, we conclude that Americans' environmental values derive from three sources: (1) religion, whether traditional Judeo-Christian religious teaching or a more abstract feeling of spirituality; (2) anthropocentric (human-centered) values, which are predominantly utilitarian and are concerned with only those environmental changes that affect human welfare; and (3) biocentric (living-thing-centered) values, which grant nature itself intrinsic rights, particularly the rights of species to continue to exist.

We derive these three sources of values from answers in the semistructured interviews. For example, we infer that an informant bases environmental protection on religious values if he or she argues that species should be protected "because God created them." We infer anthropocentric values when an informant says "If we cut down all the rainforests . . . we're going to lose out on a lot of chemicals and drugs," and we infer biocentric values when an informant says "nature has intrinsic worth, apart from its human use."

Our three categories of values overlap and encompass poorly defined subcategories. For example, one could argue that our category of religion is not logically separate, because religion includes within it anthropocentric, biocentric, and other values. And a particular anthropocentric value we found—responsibility to ones' descendants—emerged so strongly that we feel it is too hidden by our treatment of it as a subcategory of anthropocentric, although we preserve that organization because it is logically appropriate. We do not have a strong theoretical commitment to organizing the values we found into precisely these three categories but find them useful for ordering our presentation of findings.

Other researchers have identified slightly different sets of values for environmental protection. For example, both Merchant (1992) and Stern and colleagues (Stern and Dietz 1994; Stern, Dietz, and Kalof 1993) postulate three bases for environmental values: the self, other people, and the biosphere. In other words, these researchers have independently arrived at our biocentric value, they have split our anthropocentric value into self and other (a position we are neutral on), and they have omitted religion or spirituality as a basis. One reason that many scholars have overlooked religion as a basis for environmentalism may be that their initial list of values to study typically derives from environmental literature. By contrast, we asked forty-six people open-ended questions such as why they thought protecting the environment was important, a procedure that allows unexpected responses to emerge.

This chapter covers each of the three major sources that people gave for their environmental values: God, humanity, and nature itself.

Religious and Spiritual Values

Religious Models and Values for Nature

To read the literature on religion and the environment, one might not expect devout Christians to be environmentalists. In a classic essay, Lynn White argued that the Judeo-Christian view of creation is intrinsically anthropocentric, considering humanity to have "transcendence of, and rightful mastery over, nature" (White 1967). For example, an exploitative view is expressed in Genesis:

> And God said unto them, Be fruitful and multiply, and replenish the earth, and subdue it; and have dominion over the fish of the sea, and over the fowl of the air, and over every living thing that moveth upon the earth. (Gen. 1:28)

White's views have generated much criticism, both in the form of prescriptive arguments that humanity *should* consider nature to have intrinsic value (Berry 1988; Rolston 1988), and in descriptive claims that the intrinsic rights of nature have ample roots in Western culture (Passmore 1974; Nash 1989). The question we address here is, regardless of religious writings or doctrine, how do ordinary people link the environment with their systems of values?

Many of our informants draw on religious concepts to describe nature, including those who adhere to a traditional religion and those who do not. In the most modest form, these beliefs are simple extensions from the biblical account of God as the creator. Pervis, a Methodist, expresses a typical view: "I was taught God put earth here for us." However, Pervis does not see Genesis as justifying human environmental destruction.

> The Bible tells us that we have a right to use natural resources, but we have responsibilities too. *Did God give us those responsibilities along with the right to use the resources?* He gave us a brain. We should know right from wrong.—Pervis (coal mine wireman)

Pervis seems to argue that environmental destruction is wrong in God's eyes—that our God-given right to use resources is accompanied by responsibilities.

Peter, a Baptist active in his church, responds to our briefing about global warming by arguing that humans should not cause it. He uses a concept similar to "God's plan" without explicitly invoking the Deity.

[Global warming] is not a natural happening. It's something that's been brought about by man, and I think that's wrong. It doesn't fit into the big scheme of things, meaning the way that nature intended for things to happen.—Peter (logging contractor)

Gerard, a legislative council and active Catholic, more specifically refers to (God's) plan:

If a species is becoming extinct for a reason that's not related to our development, or whatever, that may be part of the plan. If there is a plan. But if we're the ones that are prompting it, then I think that it's a whole different situation. —Gerard (legislative counsel)

His reference to "the plan" clearly evokes Judeo-Christian teachings about God's divine plan for creation. To him, the plan creates a moral distinction between human-caused and other species extinctions.

A more striking example comes from Marge, an environmentalist who reported "no religious background." During the interview, she said "it doesn't seem right to kill off species" and "they have as much right to be on this Earth as we do." When we pressed her repeatedly to justify this, she finally answered that it "doesn't seem right."

Um, just because they're here. I mean, I don't know. Because God put them on this earth, so He must have had some reason [said with markedly different intonation]. I don't know. . . . *Were you being tongue-in-cheek when you said "because God put them on the earth?"* No, not really, I mean I don't really believe in God in the sense that I go to church. . . . I definitely have the scientific view of how we got here, but . . . I don't see why we should be able to decide what species live and which die. I don't see what makes us superior. I just feel deep inside that they have as much right to be here as we do.—Marge (lawyer, environmentalist)

Marge's different intonation for "because God put them on this earth" suggests that her phrase is a last resort. When pressed by our continued questioning, she was struggling to explain in words a moral value that she simply "feels deep inside." We find it striking that even this individual, who would not invoke God in other contexts, does so in order to talk about (and think about) the meaning she gives to nature. It appears that religious discourse can be useful to scaffold moral arguments even among the agnostic.

Cindy also linked ecological principles to religious beliefs. She first joked that losing some animals, like snakes, "wouldn't bother me." But she became more serious when we pursued the question, asking:

How about other animals that we don't particularly depend on, like squirrels or maybe some animal that's out in the Midwest that people just don't see that often? They're all here for a reason [pause] they gotta be food for something. [laugh] . . . *could you elaborate?* Well, if they're food for a certain animal and they become extinct then they'll try to get something else. Maybe us! [laugh] . . . It's just, it'll unbalance everything. My daughter asked me the other day, "Why did God make mosquitos?", and I said, "Well, the birds need food." She said "But Mommy, they suck our blood!" [laugh] And I said, "Well, you'll have to ask Him that when you get there."—Cindy (housewife)

Cindy's belief in the divine purpose of creation is intertwined with the cultural model of species interdependence we described in chapter 3.

Our survey results reinforce the relationship between religion and environmental values found in our interviews. We find a substantial majority agreeing with a statement justifying environmental protection by explicitly invoking God as the creator, with striking uniformity across subgroups.

58 Because God created the natural world, it is wrong to abuse it.

Earth First!	Sierra Club	Public	Dry cleaners	Sawmill workers
76	79	78	69	78

This widespread agreement with creation as a reason for environmental protection is not because our sample uncritically adopts any biblical concept. A majority of the same sample rejected the apocalyptic view described in Revelation.

61 Maybe global climate change is a fulfillment of the biblical prophesy that the world will end in fire.

Earth First!	Sierra Club	Public	Dry cleaners	Sawmill workers
33	7	20	33	26

Agreement with statement 58 about creation was correlated with religious belief, whereas agreement with statement 61 about the apocalypse was not. Respondents were asked for two measures of religious belief: whether they felt there was a spiritual force in the universe and whether they belonged to an organized religion. By either measure, those who were religious were significantly more likely to agree with our survey statement 58 and to agree more strongly. That much is not surprising.

The converse is of greater interest: of those who did not belong to any organized religion, 69 percent agreed with statement 58 about creation. Of those who stated that they did not even believe there was a spiritual force in the universe, 46 percent agreed with it.

What is going on here? Why should so many nonbelievers argue on the basis of God's creation? The earlier quotation in which Marge justified the rights of other species "because God put them on this earth" helps make sense of this seemingly contradictory response. It seems that divine creation is the closest concept American culture provides to express the sacredness of nature. Regardless of whether one actually believes in biblical Creation, it is the best vehicle we have to express this value.

A Spirituality in Nature

For other informants, the link to religion is not by way of explicit church teachings but rather through feelings of spirituality in nature. Most reported experiencing this spiritual feeling directly from contact with nature.

We used the question "*Do you feel you have any spiritual basis for your beliefs about the relationship between humans and nature?*" We did not further define "spiritual," to let informants decide for themselves whether they felt this term applied to them. Some unambiguously answer this question "No." About half the people who invoke a spirituality in nature are active environmentalists. Some people invoking spirituality are involved in traditional churches, some describe themselves as agnostic.

Catherine, a practicing Congregationalist, makes the most explicit and direct connections between God and nature.

> You're never nearer the Lord than when you're outdoors . . . out in the country or at night with the stars shining down in the pure air and so forth. . . . The most restful and calming and healing process a human being can [undergo] is to be out in the natural, quiet place of nature. . . . Nature and human beings should be in perfect harmony. It just gives you an ethereal feeling to look into the starlight or to look at a sunset or see a flower that's perfect or something. It's just unbelievable what nature can do and how it can revitalize everything in you.—Catherine (retired science teacher, environmentalist)

Catherine sees nature as God's creation and the natural world as a vehicle for humans to experience God's presence, peace, and healing energy. She has a fairly traditional Christian belief system. Yet we read Catherine's statement as expressing feelings that are derived more directly through contact with nature than through religious texts or preaching.

James, owner of a small farm and a practicing Nazarene, expresses much the same sentiment.

God has put us on this earth and has produced so many marvelous things in the natural world for us to enjoy. [Taking care of what He has given us] to me is part of, well, you could say, Christianity. . . . We tend to neglect these things that God has developed for us.—James (farmer, custodian)

Both Catherine and James readily responded to the question about having any "spiritual basis" for their environmental beliefs.

Other informants do not subscribe to traditional religious beliefs, yet they also feel a spiritual connection with nature. Abby, who says she is "sort of pantheistic," defines the spiritual as "something greater than the sum of your parts." She describes her spiritual connection with nature as follows: "I think that people and living things somehow are united by some sort of force or some sort of awareness or consciousness." Abby cites as evidence an experience alone in the jungle in Mexico, when she felt a "presence of other intelligences" (described in more detail in chapter 7).

Emma, who describes herself as an agnostic, nonetheless defines her relationship to nature partially in spiritual terms.

So I am somebody that goes walking in the woods for spiritual enlivenment and all of that. So I think there's two sides to the way I feel about it—that both my bodily existence and my spiritual existence would be hurt if the environment sustained much more of the kind of damage that it seems to be sustaining now.—Emma (coal loading machine operator)

Mark, who was raised by agnostic parents and says he has "no personal theology," was asked if he has a spiritual basis for his feelings of how humans should treat nature. He initially seems to deny it.

No, I'm eclectic, and I kind of, I'm more driven by this broad sense . . . it's an evolution, it's gone on for twenty years. And I've looked and read those people who best speak to my impulses and who have articulated best for me. It doesn't matter whether they're Buddhist or Catholic or Judaic. . . . I haven't spoken of this for years. . . . [I have] a deep spiritual sense. . . . Buddhism has been said to be the religion before religions. . . . to me it resonates because I'm always, I'm very conscious, I've come to appreciate that it's not the good people versus the bad people; it's all of us in our daily actions have a collective impact on the planet. So I've tried to slow down and look at the moment I rise to the moment I go to bed, I may be making fifty to a hundred decisions each day. But they're going to have a small, incremental effect on the planet. . . . Tic Nat Han [Buddhist writer] talks about, you bring out the spiritual vitality by being aware of how your decision-making process is going on. And so, to me, that's spiritual. You have a very spiritual sense just [in] your daily work life. . . . If you talk to my stepchildren, they would say I talk about dish washing as much [as] an environmental issue as I try to talk to them about [its] being a meditation exercise.—Mark (legislative aide)

The passages in this section show a diversity of religious systems coexisting with a shared awe, reverence, or respect for nature. For some, explicit religious teachings or God's name were invoked. For example, various informants told us that God put things on this earth, that we have an obligation to use God's gifts wisely and not damage them, or that God has a purpose for each creature. There is something compelling about the metaphor of "God's creation," as we found that even agnostics explain their environmental values with reference to it.

Others reported a feeling of spirituality in nature, which they directly experienced. For some it was related to traditional religion, for others spirituality was independent of religion. Deep spiritual feelings were expressed more frequently, but not exclusively, among those who had committed major portions of their time to work on environmental issues. Unfortunately, we inadvertently failed to include a statement about spirituality in nature in the survey, so we cannot compare survey results with the semistructured interviews.

In light of our findings, we reconsider White's classic essay (1967) in which he blamed environmental destruction on the Judeo-Christian notion of man's dominion over nature. White predicted that environmental problems would not be solved until Western society changes its religious beliefs, and he proposed an alternative religious model—the Franciscan notion of equality of all creatures—as a goal for society to move toward. However, our interviews show that Americans already use a broad range of religious teachings to justify environmental protection. Religion seems not to be getting in the way of environmental support, but instead is reinforcing and justifying it.

Anthropocentric Values

This section describes anthropocentric values, that is, values based on human benefits or human goals. The emotionally strongest of these anthropocentric values was a concern for one's descendants. Other anthropocentric values involved material utility or aesthetic utility to humans.

Descendants and Future Generations

We found that our informants' descendants loom large in their thinking about environmental issues. Although our initial set of questions never asked about children, seventeen of the twenty lay informants themselves brought up children or future generations as a justification for environmental protection.[1] Such a high proportion of respondents mentioning the same topic is unusual in answering an open-ended question. In fact, concern for the future of children and descendants emerged as one of the strongest values in the interviews. For many, this was most concretely expressed in terms of one's own children.

> *In protecting the environment, [what] are your main responsibilities?* . . . My kids. *That's the most important thing?* In my life. Kids number one, my wife's number two. Then everything else falls into place.—Pervis (coal mine wireman)

Some also volunteered that they felt strongly about future generations, not just their own children. For example, James and Peter both referred to their own children, but also clearly extended beyond them.

> The flowers that we have, and the birds, and the wildlife . . . I realize more as I've grown older that we've got to take care of it because the future generations are gonna have this . . . not only my children, but all children. For everybody, because if we continue the way [we] are, we're not gonna have these things for them.—James (farmer, custodian)

> *Would you say that protecting the environment is important?* Oh, very much so. *OK, why?* Well, because, you know, I've got a family, I mean, . . . we're leaving this world to the next generation and they're leaving it to that generation. I mean, it's just, I think it's very important. *So it's your kids really and the future generations?* Well, it's the future. It's the future of the world, really.—Peter (logging contractor)

Having children gave this value a personal and emotional reference point. Nevertheless, a similar value for future generations was held by people without children of their own. This was seen in statements by childless people such as Jack, who said we needed to "conserve for future generations." When asked why, Jack expressed his values in terms of human continuity, fairness versus selfishness, and a debt to an abstract "next generation."

I feel just for their procreation of life. . . . To our knowledge we are the only planet that has life on it. And I feel we owe it to the next generation to at least keep the environmental qualities that we have at the same or even improve on what we have. I don't think it's fair to say that we're living now, and the next generations be damned. . . . The past generations have had increasing qualities of lives, why should the next generation have a decreased quality of life? I think it's selfish for us to say that, "We're here, we're now, we're gonna use it." I think that's the wrong attitude to take.—Jack (toxicologist, environmentalist)

Other childless informants similarly gave abstract "next generation" reasons for environmental protection, such as: "[for] future generations to be able to function alright" (Tom), or "hopefully there's lots of people coming after us" (Nick).

The previous interview excerpts were in response to questions about the environment in general. The value of the environment for the sake of descendants came out especially strongly in discussions of global warming.

Our children have to live with it. . . . I'm very concerned what their future is going to be like and what their life is going to be like.—Jenny (social studies teacher)

We should not "do nothing" and wait to see if the greenhouse effect is coming because it's not fair to those that are going to be around to live with it.—Ronald (resort proprietor)

I guess I do have quite an interest, and actually I guess an anxiety about it [global warming] in sort of a way that, uh, here's my daughter [he gestures to his daughter] and I think to myself, "Well, gee, what kind of world is she going to be living in?" I mean, I really think about that, and a lot. In other words, what can I do to make my kid's life better besides givin' her a good home? I mean, I can do all those things and still the environment could be totally, a terrible place to live . . .—Doug (pharmaceutical scientist)

Doug's statement is interesting in that it defines parental responsibility as extending beyond the traditional realms to include environmental protection.

Furthermore, people extend their concern for descendants beyond the first generation. Several people volunteered that their concern was multigenerational (again, this was not one of our questions).

> I think when you have kids and you think about long term, you know, like we might be comfortable now, but what are they going to have when they get older? I think basically, [that is what] makes you decide what's important and what's not important that much. . . . It might not happen for another hundred years, but maybe not for them, but their kids . . .—Cindy (housewife)

> [Informant is discussing people whom he feels are not sufficiently concerned about the environment.] They're complacent, I guess. They figure that "It ain't gonna be in my lifetime.". . . [pause] They don't figure that, that it might be in their kids' or grandkids' or [pause] great-grandkids'. It's gonna be in somebody's lifetime, that's for sure. —Walt (retired machinist)

> It [global warming] probably won't affect me personally. I mean, I don't have that much longer to live. You know, thirty years down the road I'm going to be gone probably. I'm almost fifty. But the kids, my kids, and their kids, it will be a problem. In the not too foreseeable future, you know, it could be catastrophic . . . unless they do something and do it now.—Jenny (social studies teacher)

The above quotations suggest a value for environmental protection that stretches far into the future. This conclusion needs to be validated by other researchers because it is especially important for long-term questions such as climate change and species extinctions. The time perspective expressed above conflicts with the primary economic methods used to make decisions about future costs and benefits, as we describe in the next section.

The value of the environment for one's descendants appears to have been strong enough to stimulate some people to write their political representatives, judging from a comment by one congressional staff member.

> All of a sudden the mail starts coming into the Congress saying, "I don't want my children to have to deal with the consequences of a global climate change." . . . That . . . didn't exist three years ago.—Sam (congressional staff)

Concern for children also motivates higher levels of action by some environmentalists. Margaret says that environmental activism is "hard work," with little "visible feedback for the effort." She cites her children as motivation to continue and describes two reasons, one based on an optimistic and one on a pessimistic outcome.

> I feel a need to set an example for my children. Suppose we have enough time that when their generation comes along there's still time to do the kind of work that I'm doing. I want them to have a sense of their responsibility in doing that. The other thing is that if there isn't the time for their generation to continue on being the stewards of this planet, then I want to rest easy in the notion that I did what I could to provide what I could for my children. And that doesn't mean their own bedroom each. . . . I think it's a sense of guilt when I talk about the kids. If I don't do what I feel I need to do for the kids, that the guilt will be more overwhelming than the effort to do it. . . . [and] I have a responsibility to teach these kids to take over.—Margaret (activist, environmentalist)

Margaret is not only trying to improve the world for her children but is also setting an example and, like Doug in the earlier quotation, defining her own parental responsibility partly in environmental terms. Few people devote as much of their lives to environmental work as Margaret does, but our survey respondents overwhelmingly agree with the rationale that she and Doug express.

111 Working to try to prevent environmental damage for the future is really part of being a good parent.

Earth First!	Sierra Club	Public	Dry cleaners	Sawmill workers
97	100	93	97	85

We tried to gauge the relative strength of family and environment by opposing them in a survey question.

63 My first duty is to feed my family. The environment and anything else has to come after that.

Earth First!	Sierra Club	Public	Dry cleaners	Sawmill workers
13	73	70	73	70

This is one of the questions that really separates out Earth First! members, with almost 90 percent of them apparently putting the environment ahead of family.[2] We find more surprising that only 70 percent of the other groups agree with this statement, suggesting that 30 percent of the population puts the environment ahead of feeding their family. (They may in fact be disagreeing with the presupposition of our statement, believing that environmental protection is compatible with, or necessary for, feeding their family.)

If some adults feel a responsibility to pass on an intact planet, it raises the question of whether the receiving children feel the same thing. We interviewed only adults, but we report observations from our informant

Margaret, an organizer who had worked with high school students. She recalls her initial contacts with their environmental clubs, as part of Earth Day preparations.

When we went into the high schools . . . what I noticed is that these high school kids really have a sense of rage, and anger. They really do believe that what they are being given is so much less than what their parents were given, in terms of the planet. And that people [adults] have really blown it. That people have not done what they are supposed to do. And they [kids] aren't getting what they should have been handed over. . . . And they're angry about the environment. They really feel they've been given the short end of the stick here. . . . And there's a real lack of trust of adults because of that.—Margaret (activist, environmentalist)

Based on the interviews, we identify a value given to future generations in general, and to one's descendants in particular, as one of Americans' most widely and strongly held points of reference for environmental values. One could object that this could have been exaggerated in the semistructured interviews, since the latter part of those interviews focused on long-term global environmental problems. Nevertheless, the finding is validated on the survey.

25 We aren't justified in using resources to benefit only the current generation, if that creates problems for future generations.

Earth First!	Sierra Club	Public	Dry cleaners	Sawmill workers
100	100	93	90	82

Furthermore, this long-term perspective shows no significant erosion when environmental protection is claimed to require sacrifices today.

24 We have to protect the environment for our children, and for our grandchildren, even if it means reducing our standard of living today.

Earth First!	Sierra Club	Public	Dry cleaners	Sawmill workers
100	100	97	87	74

These survey questions verify our sense from the semistructured interviews that these values are widely held in the culture.

Implications of the Value for Descendants and the Future

We relate our finding of the value of descendants and the future to two quite different theoretical issues. The first is the economic method of "discounting the future," which is often applied to environmental problems. The second is the philosophical question of why people might care strongly about the future.

First the economic question. The greatest damage predicted from climate change would occur decades or even centuries in the future. The discipline of economics holds that if costs incurred in the future are to be used in decisions today, they must be discounted to their present value.[3] For example, the Bush White House argued as follows, with regard to estimates of damages from global warming in the future:

> Using a 5% real interest rate, a global [agricultural] loss of $170 billion in 2050 amounts to about $9 billion in 1990 dollars. Thus . . . [trade problems costing $35 billion today are] estimated to be more important in economic terms than even pessimistic estimates of the effects of global warming, largely because the former must be borne in the present and the latter may occur, if at all, in the relatively distant future. (Council of Economic Advisors 1990)

These kinds of economic calculations make future costs appear ever smaller, the further into the future the damage occurs. Discounting is a well-established principle in economic analysis, essential for problems like investment evaluation and corporate planning. Consumers apply it, without knowing its name and perhaps in imperfect form, to purchases and investments. However, some scholars argue that the concept of discounting should not apply to long time periods, transfers across generations, or to the environment, and others would dispute the appropriate discount rate (e.g., Redclift 1987, 39; Norgaard and Howarth 1991).

Our informant's views on children and future generations, especially in the multigenerational value statements quoted in the previous section, suggest to us that discounting may be inappropriate for another reason. We find that people place great value on the environment that their children and grandchildren will inhabit, and that value does not seem to diminish for succeeding generations. In other words, discounting would violate basic values about the environment, our descendants, and the continuity of society. Most citizens do not seem to discount the future when making their own decisions about long-term environmental issues.

Our second question is, why does a responsibility to descendants provide such a strong basis for environmental values? The most thorough arguments come from philosophers (e.g., MacLean and Brown 1983). One standard way of describing responsibility to descendants is to call it *intergenerational ethics,* an obligation between our generation and future ones. Such an ethical concern can be seen in some of the quotations in this chapter. MacLean (1983) takes another approach. To resolve problems with arguments that benefits should accrue to future generations, MacLean argues that meaning, as well as happiness, determines the quality of life. Thus he argues that "a commitment to securing resources and opportunities for future generations . . . is an appropriate way of expressing our belief that the society and culture that matter to us are important enough to survive into the future" (MacLean 1983, 194). In other words, knowing that the society will endure makes our lives more meaningful and thus of higher quality today.

We can see this type of argument in a quotation of Tara "you take pride in history . . . the way it's been and should continue" as well as in the following:

> It's just as suicidal to do something that's gonna affect two generations away or other life on the planet as it is to do something that will kill you personally. . . . I don't think it really matters whether you have your own child or whether as a . . . human being you're connected to everybody in a certain familial way. . . . I think there's a continuum of life that goes beyond our time.—Abby (shop owner, environmentalist)

While people may relate most emotionally to their own descendants, the quotations from Abby and Tara, as well as earlier quotations by Jack, Nick, and Tom, all childless, reveal a value for continuing the broader human tradition.

The desire to protect the environment for our descendants appears to be a nearly universal American value. For many of those with children, the semistructured interviews revealed that this value also has strong emotional content. Although MacLean (1983) offers one plausible explanation for this value, our research did not attempt to confirm or deny his hypothesis. Our intuition is that the academic discourse about intergenerational ethics and the endurance of society has captured part of the value for the future we see in our informants, but does not fully explain its emotional force. Given the importance of this value, we find it curious that this value has not played a more central role in public discussion and scholarly investigation regarding global environmental change.

Utilitarian Arguments

We use the word *utilitarian* for situations when the value of something is proportional to how useful it is to humans.[4] White's essay (1967) predicted that Americans would regard nature in utilitarian terms, due to the Judeo-Christian dictate in Genesis that humanity should have mastery over nature. If this prediction is correct, then we would expect our interviews to be filled with utilitarian arguments for environmental protection and little else. Further, the faithful among our informants would tell us that man's "natural" position is not as a subset of nature, but as a manager of nature. In fact, very few take this position. One of the few, Ronald, expresses its most extreme form.

> The Creator intended that nature be used by man. I wouldn't agree with the Indians that nature is there to be worshipped. Theirs was a nonprogressive society compared with the white man's. Their society didn't change over centuries. . . . I don't think the environment stands apart from man. I think that it is there to serve man, and if man sees that what he's doing is ultimately against his best interest, then he's obviously got to take [steps to correct it,] but I don't think the environment is a separate entity that has any kind of right to exist. I think it is there for man's best interest. At least that's the way I see the scheme of things as far as the Creator, spiritual aspect of it goes.—Ronald (resort proprietor)

This purely utilitarian view, far from being the predominant response, is shown by our survey to be held by only a small minority. Contrary to White's prediction, Americans overwhelmingly reject statements that nature's only function is to serve man.

80 Plants and animals are there to serve humans. They don't have any rights in themselves.

Earth First!	Sierra Club	Public	Dry cleaners	Sawmill workers
0	7	23	10	31

75 People's only responsibility to nature is to make it serve their own best interests.

Earth First!	Sierra Club	Public	Dry cleaners	Sawmill workers
0	0	24	17	11

Our survey result is also consistent with that of a U.S. national probability sample (Dunlap, Gallup, and Gallup 1993), in which 69 percent agreed that "Plants and animals do not exist primarily to be used by humans."

Ronald argued for the Genesis view by stating that nature should not be "worshipped" by humans, seemingly an obvious statement that members of any mainstream American religion would agree with. Yet, even when our survey employed this strong language, opposing use to nature worship, majorities of three of the five groups disagreed.

69 The Creator intended that nature be used by humans, not worshipped by them.

Earth First!	Sierra Club	Public	Dry cleaners	Sawmill workers
0	30	35	52	59

These survey results buttress our earlier conclusion that the ways that ordinary people think about the environment and the Creator are not consistent with the human utility view of Genesis. Nevertheless, less extreme utilitarian arguments (ones that allow for utility as one of many reasons for environmental protection), are commonly expressed by our informants. For example, consider two of the cultural models discussed in chapter 3: First, the model that humans are a part of and dependent on nature leads directly to utilitarian arguments to protect nature. This was evident in many of the quotations cited previously about human dependence on nature. Second, the model of chain reactions further strengthens those utilitarian arguments.

More specific utilitarian arguments were raised when we asked why people are concerned about species extinctions. The most common reasons are utilitarian or based on the interdependency argument mentioned in the previous chapter. For example, Doug, who is environmentally well informed and concerned, justifies species preservation in utilitarian terms.

Probably also plants and animals extinctions would probably be very important too. *How's that? It wouldn't affect you directly like food costs. . . .* let's say we're dealing with six thousand known species. We're startin' to really lose a lot of major promises, especially plants. You talk about deforestation, for one thing. There might be certain plants that might be cures for cancer. [But] because, hey, we gotta make farmland, or our third world country's growin', we don't care about this particular plant. And, because that species of plant no longer exists, you just probably destroyed a cure for some known disease. And also, with animals [I'm concerned about animal extinctions], being an animal lover and all that stuff.—Doug (pharmaceutical scientist)

Even the environmentalists typically make utilitarian arguments (although they are different from most others in also volunteering non-utilitarian arguments).

> Also [I'd] . . . just like to elaborate on destruction of the tropical rainforest. . . . There are potentially thousands of plants that are unclassified and unknown to man right now that might have possible medical benefits or other type of benefits to man that are undiscovered as of yet. We might lose them . . .—Jack (toxicologist, environmentalist)

> People say that if we cut down all the rainforests and destroy all of the thousands of species which are slowly becoming extinct, we're going to lose out on a lot of things. Apparently there's a lot of species out there which we use for different chemicals and drugs, all sorts of stuff that we can't live without.—Marge (lawyer, environmentalist)

These environmentalists' prominent mention of utilitarianism is consistent with an observation by Tribe (1974), who notes that while most environmentalists base their views on concern for nature itself, they often rationalize these concerns on a utilitarian basis. The strategy of using such arguments to generate public support for environmental causes may be based on a belief that utilitarian arguments appeal to the broadest spectrum of people.

For example, the medicinal uses argument has been made by environmental groups arguing for the protection of species in general and tropical rainforest species in particular.

98 There are probably thousands of medicinal and other useful plants that are unknown to science that we might lose because of global climate change.

Earth First!	Sierra Club	Public	Dry cleaners	Sawmill workers
100	93	83	93	63

Whether or not this is factually correct, the high agreement levels shown in our survey demonstrate that it has been communicated effectively to the public.[5]

Aesthetic Arguments

Some of our informants argued for environmental protection based on aesthetic enjoyment of the environment (strictly speaking, this is another form of utilitarianism). For example, when asked to elaborate on her spiritual connection with nature, Emma says:

I guess it's easier . . . to start with the aesthetic of the spiritual side. I feel healthy in the presence of complex life or complex anything, I guess. We're talking about art, for instance. I respond . . . less to geometry and more to complex shapes and colors. And I think that comes from having grown up in the woods and just associating that with all things pleasant, and secure, and calm . . .—Emma (coal loading machine operator)

In our survey, most agreed with even an extreme statement that beauty resides in nature.

121 Nature is inherently beautiful. When we see ugliness in the environment, it's caused by humans.

Earth First!	Sierra Club	Public	Dry cleaners	Sawmill workers
77	63	70	70	33

Another human use of the environment is to calm or "revitalize" oneself. This can derive either from aesthetics or from the feeling of spirituality in nature discussed earlier. Only a few of our informants volunteered this view, such as Emma and Catherine. Yet, when we asked about it in our survey, solid majorities of all groups agreed.

101 Being out in nature can revitalize everything in you.

Earth First!	Sierra Club	Public	Dry cleaners	Sawmill workers
93	89	93	77	78

Aesthetics can be one type of utilitarian value. This argument is made by Steve.

I believe the world's here to serve man . . . though there'd be no reason to chop down a four-hundred-year-old tree in my mind, because it can make fifty thousand picnic tables. You know, there is such a thing as beauty just by itself in nature. . . . Just because looking at it is one way it serves man [or] being able to walk through the forest. That's a service, it isn't just building things or processing things.—Steve (auto fuel efficiency planner)

People also used the value of aesthetics and the enjoyment of observation to justify preserving species. In this context, it is natural for them to relate species extinctions most directly to animals they had personally seen in their neighborhoods or in zoos.

I'm an animal lover. . . . You always feel sorry . . . when you hear about them becoming extinct, you know like in Africa where they're killing the elephants for the ivory and all like that. You know, it's terrible. . . . You need birds around. You can hear 'em singing, it makes you feel better, you know. [laugh]

I used to have a bird feeder. . . . Before they built all that [points to townhouses], I used to hear the tree toads, you know, in the spring. . . .—Jane (retired insurance actuary)

This aesthetic enjoyment in part accounts for our sense that public discussions of species preservation most often cite large mammals—not rodents, insects, or lichen. This is sometimes reflected in law, as in the Marine Mammal Protection Act, which gives mammals such as whales, dolphins, and seals greater protection than fish, some of which are equally endangered.[6]

Over a third of the lay informants volunteered that they had a personal interest or empathy with animals or described themselves as an "animal lover." This was an unexpected category sometimes volunteered in the semistructured interviews to explain concern for other species. We thus added this to the survey, asking whether respondents would agree with the statement, "I consider myself an animal lover." To our surprise, 97 percent of our survey respondents did.

Biocentric Values

This section explores the rationales for environmental protection based on the value or rights of nature itself. Such an ethic has been hypothesized to explain the increasing U.S. environmentalism of the sixties. In a seminal article, Heberlein (1972) saw a developing "land ethic," a concept derived from Leopold: "A thing is right when it tends to preserve the integrity, community, and beauty of the natural environment. It is wrong when it tends otherwise" (Leopold 1949). Heberlein attributes this development in American ethics to increases in two factors: public awareness of consequences of pollution and attribution of responsibility for environmental destruction to specific individuals.[7] More recent writers refer to the land ethic as a "biocentric" view, to reflect that ethics derive not just from humanity or from God, but from nature itself (Devall and Sessions 1985; Naess 1989). This section covers three variants of biocentric positions: humans should not harm nature because we are part of nature; species have a right to continue; and nature has intrinsic rights broader than mere species survival.

Humanity as Part of Nature

Some biocentric value statements derive from the assertion that humans are part of nature and therefore have the same fundamental capabilities and are subject to the same ethics.

We're very much a part of [the environment.] And we have a big tendency to forget that. . . . You just have to remember that we're all just a bunch of monkeys. And when we start expecting ourselves to behave too much differently than a bunch of monkeys, we'll just get frustrated. . . . We're enough different from some of the other critters that we tend to forget how similar we are. We get all mixed up about that . . . and just because they can't talk our language back, we just think that there's a more qualitative difference.—Emma (coal loading machine operator)

Humans are a part of nature. . . . There should obviously be a harmony, a balance between [them].—Ronald (resort proprietor)[8]

The view that "humans are part of nature" and the injunction against human meddling with nature are widely agreed with in the survey.

77 Humans should recognize they are part of nature and shouldn't try to control or manipulate it.

Earth First!	Sierra Club	Public	Dry cleaners	Sawmill workers
97	74	72	57	69

The question of human control of nature can be separated into two parts: whether humans have a right to control nature, and whether they are able to do so. The two parts have been separated in other surveys. The first part is verified as a majority opinion in Dunlap, Gallup, and Gallup's (1993) U.S. national sample. They found that a 44 percent minority agreed that "Humans were created to rule over the rest of nature," whereas a 51 percent majority disagreed. The second part of statement 77, that humans cannot successfully manipulate nature, has been tested in a survey taken in the State of Washington. In that survey, pairs of statements were opposed. To the statement "Despite our special abilities, humans are subject to the laws of nature like other species," 62 percent agreed. The alternative choice was "Because we are human, we are not subject to the laws of nature as are other species," to which 20 percent agreed (Olsen, Lodwick, and Dunlap 1992, 66).

The idea of humans as part of nature is also used, less frequently, to justify exploitation of nature. Bert, a hunting guide, was quoted earlier as saying that nature is for animals as much as it is for people. Nevertheless, he encourages the active human management of nature, and sees hunting as intrinsic to humanity.

> We do a better job [of managing wildlife] now than we did a hundred years ago or even fifty years ago. . . . *So wildlife would be one of the things that man has some kind of a right to use . . . ?* Well, I think so . . . of course, there's people that are against hunting and so on. . . . But . . . man's a predator just like a wildcat or a bear is, or a coyote or anything else. I mean, man was born being a predator. That's how man got to be here is by killing squirrels and rabbits or whatever other thing. . . . Man doesn't have to kill things like they used to, but I still think that's a part of it . . . you know, if man wasn't supposed to kill something he wouldn't have been taught that to begin with.—Bert (resort proprietor, hunting guide)

Bert uses two lines of thought to support his views on hunting. Man has a responsibility to manage the wildlife population, and he also has the right to kill animals because he is naturally a predator. Bert's perceived right to kill is less of a God-given right than an intuitive animal instinct. Nonetheless, as a part of nature, this logic would grant humanity a right to both manage wildlife and to kill individual animals.

Hunting is an issue that divides our subgroups.

114 There is nothing wrong with killing individual animals, as in hunting, as long as you don't kill so many that you threaten the population.

Earth First!	Sierra Club	Public	Dry cleaners	Sawmill workers
40	67	43	43	89

This distribution does not fit the idea of groups arranged in order of environmental sympathies, but it does make sense in terms of the particularities of individual groups. We infer that Sierra Club members have accepted the value of species preservation, making it easier for them to accept killing of individual animals than does the public. Although Earth First! also has a strong species preservation ethic, their deviation from the Sierra Club on this question may be due to a counterbalancing strong representation of animal rights activists in Earth First!. We believe that so many sawmill workers agree because, as rural blue-collar workers, this subgroup undoubtedly includes many hunters.

Whether or not they consider it acceptable to kill individual animals, solid majorities in every group acknowledge that hunters can consistently be environmentalists.

120 It's not inconsistent to be a hunter and a dedicated environmentalist as well.

Earth First!	Sierra Club	Public	Dry cleaners	Sawmill workers
74	67	73	60	82

Rights of Species to Continue

In our interviews, the value of nature was most often voiced with regard to the survival of other species. For example, little concern was expressed over sea level rise because rocks would be submerged, or about ozone depletion because we would grieve the loss of the ozone itself. Rather, when environmental changes were considered wrong, it was due to their impact on living creatures.

The principle of species preservation is already recognized in both U.S. law (the Endangered Species Act) and international law (the Convention on International Trade in Endangered Species). We address the independent question of how strongly, and with what logic, citizens support the values implied in these laws.

Informants cited multiple values leading them to want to prevent species extinctions.

Plant and animal life becoming extinct would concern me because that's something you can't replace.—Walt (retired machinist)

Extinction is so permanent. . . . I was talking earlier about obligations. I think we do have an obligation to try to preserve our species [of plants and animals] from extinction if we can.[9] *Is that an obligation to voters, to those species, to nature, . . .?* To nature. To the whole ecosystem. To the reason we're all here.—Gerard (legislative counsel)

Some of the environmentalists used more explicit language. For example, Marie talks about a right to evolve and also distinguishes natural from human-caused extinctions.

I don't mind natural extinctions, but I'm not too enthusiastic about extinctions that are directly caused by man. I feel that a species has a right survive and be able to survive on its own and be able to change and evolve without the influence of whatever man does. I don't want to see man kill [any species]. If it's going to happen, it should happen naturally, not through anything that man has an influence on.—Marie (health inspector, environmentalist)

Marie adds, "once a species is gone, it's gone for good." A special moral status for human-caused extinctions is also suggested by Nick.

As for extinction, . . . that one's a hard call too. . . . I see the world as a place of constant flux, so there's always things coming and going. I don't like to be responsible for the ones going.—Nick (small manufacturing plant owner)

On the other hand, several informants expressed reservations about how much species extinctions should count against other human concerns. For example, Wilbur, when asked why he said he was concerned about species loss, began by justifying species protection in terms of the value for descendants described earlier. This anthropocentric value, naturally enough, does not lead him to give an absolute moral status to extinctions.

I really want to keep this world, the earth, alive for the kids. I hate to think our kids would never be able to see the things [the species] we've seen in our lifetimes. But I can see both sides, you know. If it isn't something we use, I'm not sure I'd feel that strongly. . . . It's really a pain to worry about something that may never happen or something that we may not be able to do anything about. . . . If you look very carefully at the insect population, every little thing affects their lifestyle [*sic*]. You've got to look very closely at the cycle, it's so easy for us to just screw it up. . . .—Wilbur (retired fireman)

Similarly, Sam says:

I don't believe in preservation of species in the sense that a lot of environmentalists do. I think that the evolutionary process envisions extinction. . . . There's a difference in that sense between an ecologist and an environmentalist, and I'm more of the ecologist that believes that the evolutionary process leads to some things becoming extinct. And the symbiotic relationship [between humans and nature] has got to recognize that.—Sam (congressional staff)

For Sam, humans are part of an "evolutionary process," so it is natural for them to cause some extinctions. He and Wilbur, although they might prefer not to cause extinctions, would weigh that value against others. For several informants cited earlier, species preservation, and especially preventing human-caused extinctions, seemed to be more absolute, not just one value to be weighed against others.

The responses to several statements on our survey show how strongly the value of species preservation is held by diverse groups of Americans. Among our samples, only the sawmill workers agree notably less frequently. This may be the result of regulatory threats to the timber industry in the Northwest in the name of preventing extinction of the spotted owl.

49 All species have a right to evolve without human interference. If extinction is going to happen, it should happen naturally, not through human actions.

Earth First!	Sierra Club	Public	Dry cleaners	Sawmill workers
100	82	87	77	59

35 Preventing species extinction should be our highest environmental priority. Once an animal or plant species becomes extinct, it is gone forever.

Earth First!	Sierra Club	Public	Dry cleaners	Sawmill workers
97	78	90	90	40

All groups (except the sawmill workers) overwhelmingly support species preservation for nonutilitarian reasons. Even given the emotional and direct job-related aspects of this issue for sawmill workers, almost half of them nevertheless support the principle.

104 If there is no economic, aesthetic, or other human use for a species, for example, some lichen out in the desert, then there is no reason to worry much about it becoming extinct.

Earth First!	Sierra Club	Public	Dry cleaners	Sawmill workers
0	15	13	17	52

The agreement level can be eroded by using language suggesting that support for species preservation is an absolutist position that does not allow for natural extinctions. Here we used a survey question with advocacy language similar to Sam's statement.

45 I don't believe in preservation of species in the way some environmentalists do. In nature, evolution includes extinction.

Earth First!	Sierra Club	Public	Dry cleaners	Sawmill workers
24	48	37	40	74

Yet even with this wording, species preservation is supported by majorities of all groups except the sawmill workers.[10]

In one other statement in our survey, we really pushed to see how strongly species extinction was valued by opposing it to human suffering and loss of human life.

84 I would rather see a few humans suffer or even be killed than to see human environmental damage cause an entire species go extinct.

Earth First!	Sierra Club	Public	Dry cleaners	Sawmill workers
90	48	43	56	22

Pitting species against human welfare reduces agreement, but still 40 to 50 percent of the moderate three subgroups would rather that humans suffer or die than cause extinction of "an entire species." Even granting that respondents may be thinking of species like pandas or whales when answering, we find this surprisingly high.

Our interviews provide no evidence of an ethic to preserve ecosystems rather than isolated species. Our lay informants seemed to think that species preservation required refraining from directly killing endangered animals, rather than by protecting their habitats as refuge areas. They have not yet understood that species are endangered not so much by the gun as by the development plan, the chainsaw, and the bulldozer. The idea of preserving ecosystems rather than species is fairly new and may take some time to percolate to the average person. It should not be difficult to convey, however, because it follows closely from the model of species interdependency, which chapter 3 already demonstrated to be widespread.

Intrinsic Rights of Nature

A final biocentric value expressed in the interviews is that nature itself has rights, including but going beyond the rights of species to survive. In the semistructured interviews, this right was volunteered most often by environmentalists.

> I think [my environmental vision] has something to do with a basic sense of justice and fairness towards everything. One of my pet peeves is that people think that when we talk about justice and fairness it should be about the human race. And I think that we need to be just as fair and just towards the plants as we do towards the people. I do eat plants, but I suppose it comes down to renewableness.—Margaret (activist, environmentalist)

> I don't just think of [species] as resources, but I think that's definitely one of their uses. But also it doesn't seem right to kill off species. I think they have as much right to be on this Earth as we do. . . . Just the fact that we have more brain power and consciousness than all these animals doesn't make us any better.—Marge (lawyer, environmentalist)

> *Would you say that people have the right to use some natural resources?* . . . I think they have some right to go on these lakes and forests. That's probably why they're here. You know, it's not just for loons [a northern U.S. bird], but probably they're there as much for loons as they are for humans.—Bert (resort proprietor, hunting guide)

Taking this idea still further, a few believe that humanity's intellect and capacity for self-reflection carry with them a greater burden of responsibility.

I take great issue with the fact that we have the right of dominion over resources. I think we are just one among many species, although we are the most self-reflexive of all species. That unique feature does not give us the right to eradicate the planet. It gives us an even greater responsibility to pause and reflect on this creation.—Mark (legislative aide)

If any species has to go, I think it should be the human species. We're the only species that's destroying the earth. Which other species is destroying the earth? Destroying their home? It's just ridiculous. We're making things unpleasant for every other species on earth almost. What right have we to do that?—Marie (health inspector, environmentalist)

To understand Americans' sense of moral obligation to other species, our survey included several questions on this topic. Because the environmentalists most often volunteered the idea of nature itself having rights, we did not expect this to be a widely shared value. To the contrary, the survey showed that when we raised the question, a large majority of our respondents did agree.

16 Justice is not just for human beings. We need to be as fair to plants and animals as we are towards people.

Earth First!	Sierra Club	Public	Dry cleaners	Sawmill workers
97	85	90	83	63

51 Our obligation to preserve nature isn't just a responsibility to other people but to the environment itself.

Earth First!	Sierra Club	Public	Dry cleaners	Sawmill workers
97	100	87	90	82

These high agreement levels drop only slightly when rights of humans are opposed to those of nature.

50 Other species have as much right to be on this earth as we do. Just because we are smarter than other animals doesn't make us better.

Earth First!	Sierra Club	Public	Dry cleaners	Sawmill workers
97	78	83	83	56

That most Americans believe nature has intrinsic rights is also supported by the Dunlap, Gallup, and Gallup (1993) poll cited earlier, in which 69 percent agreed that "Plants and animals do not exist primarily to be used by humans."

Predictably, support for species preservation drops into the minority if it is placed above human survival (the position argued in the earlier quote by Marie).

78 If any species has to become extinct as a result of human activities, it should be the human species.

Earth First!	Sierra Club	Public	Dry cleaners	Sawmill workers
80	22	21	33	7

While support dropped dramatically with this last question, it is perhaps more remarkable that a fifth of our general public sample agrees with a statement that might be attributed only to radical environmentalists like Earth First!. It appears that a cultural model of retribution or punishment is being invoked here. That is, the crime (species extinction) is so reprehensible that by committing it the person (species) responsible may forfeit their own right to continue to exist.

Conclusion

We find a diversity of bases for environmental values, which we have grouped into the broad categories of religious, anthropocentric, and biocentric. Religious bases include specific religious teachings, attribution of a spiritual force to nature, and religious metaphors applied in nonreligious contexts. Anthropocentric values include preserving the environment for our descendants (the anthropocentric value described with the strongest feeling), practical utilitarian values, and aesthetic appreciation of nature. Finally, biocentric values range from a vague feeling of oneness between humanity and nature to the idea that nature has rights and deserves justice. We use these three categories more for expository organization than as an important analytic distinction. As we noted, even informants with the strongest biocentric views also argue in anthropocentric utilitarian terms, and we see a gradation rather than a sharp dividing line between a number of the categories.

Discovering shared values does not necessarily imply that people will act on them. Nevertheless, our data demonstrate that environmental values are now closely tied to many other deep value systems in American culture. Further, environmental values that many would consider radical are held by substantial minorities: for example, even if we exclude members of our two samples of environmental groups, 30 percent of our survey respondents would not put "feeding my family" above environmental protection, and 40 to 50 percent would prefer that a few humans "suffer or even be killed" than have humans cause extinctions.

This strength and diversity of supporting values suggest that today's environmentalism is unlikely to be a passing fad. These results also render some previous explanations of public environmentalism, reviewed in chapter 1, implausible. For example, the explanations based on single causes, such as dangers being exaggerated by environmental activists or people wanting wilderness recreation, clearly are considering far too narrow a range of the cultural bases of environmentalism.[11]

Our findings support the positions of those few environmental researchers who have previously identified values as an important component of environmentalism. We would conclude that Milbrath (1984), Cotgrove (1982), Stern, Dietz, and Kalof (1993), and Stern and Dietz (1994) are justified in basing their measures of environmentalism heavily on values. Our findings also support the views of philosopher Mark Sagoff (1991), who has argued that utilitarian values are not a sufficient basis for environmental protection and that environmental activists should argue on nonutilitarian values as well. (Wilson (1992) makes a similar argument for preventing extinctions.) Sagoff's advice—at times pitched directly to activist groups—remains largely rejected, even though our data suggest a wealth of nonutilitarian values that could be tapped. Similarly, Gore justifies his personal environmental values based on his own Southern Baptist faith (1992, 242–248). This type of traditional Judeo-Christian basis has been criticized or ignored by the environmental intelligentsia (and is the converse of White's (1967) thesis), but we find that it corresponds to one strong environmental motivation held by the faithful among our informants.

6

Cultural Models and Policy Reasoning

This chapter describes American reasoning about environmental policies. We briefly cover policy reasoning by specialists. Policy reasoning by laypeople is explored in greater detail because of its practical value in understanding political support for, or opposition to, policies that may be publicly debated in the future. Lay policy thinking is also of interest because it provides a window on how people translate their beliefs and values into prescriptions for action. It is important to remember, however, that our analysis of policy preferences is not intended to identify better or worse policies, because public preferences may be based on faulty use of cultural models, ignorance, or greed.

Although some policy issues common to any environmental problem are considered, the focus is primarily on global warming. We consider policy debates about three general strategies for coping with global warming: adaptation versus prevention, act now or wait until the science is more certain, and lifestyle change versus technology. Two specific policies are also covered: automobile fuel standards and energy taxes.[1] A final section describes public views of the institutions involved with these policies: science, industry, and government.

In the semistructured interviews, policy reasoning is inferred from two different parts of the interview, which we carefully distinguish. First, before the briefing, we asked our informants what they thought should be done about the greenhouse effect without offering specific suggestions. Answers to this question are called "volunteered policies." After all our other questions had been asked, we gave informants a briefing on global warming, comparable in detail to a major article on the topic by a weekly news magazine (the interview, including the text of the briefing, appears in appendix C). Then we presented a set of specific policies and asked informants to evaluate them. Our second set of policy inferences draw from these postbriefing reactions to our suggested policies. No similar briefing was given during the fixed-form survey, so no similar before-and-after distinction is required.

For most of the specialists, the policy discussion before the briefing was not very different from afterward. For the laypeople, the two were typically very different. Some specialists responded to our interview question "*What have you heard about it* [*the greenhouse effect*]*?*" by producing most of the content of our prepared briefing. For them, we gave an abbreviated briefing covering only points they had not mentioned. We note that lay policy discussions following the briefing are not a fair representation of general public reactions to those policies, since the briefing gave informants a higher level of knowledge than the average voter. It may, however, be an approximation of voter thinking as it would unfold during extensive public debate on these issues, or, in the longer term, after more appropriate cultural models are developed for global warming.

Volunteered Policies

In the initial part of the semistructured interviews, before our briefing on global warming was read to them, we asked those informants who said they had heard of the greenhouse effect, "*Do you think the United States should do anything about this?*" and followed up with "*What?*" At this point, prior to our briefing, informants' policy suggestions would have to be based on information they obtained prior to the interview, whether from the media, their preexisting cultural models, or, in the case of the specialists, industry or research publications and conferences. We contrast the specialists with the laypeople.

Specialists' Volunteered Policies

We chose examples of policies volunteered by three specialists who rely on different information sources: a local environmental activist who draws on environmental group publications; an auto industry lobbyist who draws on his engineering experience, company reports, and his company clipping service; and a congressional staff member who draws on research publications.

The first example, an environmental activist, is an officer of his local chapter of the Sierra Club in Pennsylvania. Unlike the other two specialists we cite, he did not have a job or professional role involving policy analysis. Nevertheless, his volunteer role as an activist and his familiarity with material from the national Sierra Club office made his answer more comprehensive than that of most lay informants.

> Now this is the Sierra Club position. I would like to see increased fuel efficiency, in automobiles primarily, but also efficiency in electric generation, efficiency in transportation, efficiency in the way we grow and produce food . . . increased efficiency in electrical appliances. . . . *Any other things besides fuel efficiency?* Resource recovery. I'd like to see a lot more aggressive recycling programs. . . . I'd like to see a lot better mass transportation. I mean, right now the American mass transportation system is absolutely dismal.—Jack (toxicologist, environmentalist)

Although Jack did not mention other greenhouse gases such as CFCs, this may have been deliberate. We infer this from his comment that our briefing was "very thorough . . . you mentioned CFCs, which are . . . minor . . ."

The industry specialists each focused on policies affecting their own industry. We discuss one example here—Andy, a "government liaison" for a large American automobile manufacturer. Andy has a technical degree and describes his work as "I interface between [my company] and the government on alternative fuel issues." He had also worked previously for the company on fuel economy regulations.

> *What kind of measures would you visualize being done?* The first thing is what we're doing already with eliminating the CFCs, particularly like from air conditioning and some of the plastics that we're using. . . . Now what I'm afraid of is that we react in ways that are actually counterproductive. Some of the proposals that I read about doubling the average fuel economy of a vehicle on a miles per gallon basis. . . . If you've got a five-person family that no longer can get a five-passenger vehicle, they now have to get two vehicles. Haven't we lost somewhere in the process?—Andy (auto lobbyist)

He went on quite a while about problems he saw with automobile fuel efficiency standards. When the interviewer pushed him on other possible ways to deal with global warming, apart from automobiles, he asked to "stay on the auto for a moment longer" and discussed alternative automobile fuels. When asked again if he could suggest other (nonautomobile) measures, he said "Not that I can think of at the moment." His knowledge of the effects of global warming was not much better than that of the general public; effects he mentioned were melting arctic ice, sea level rise, and skin cancer.

As someone who is supposed to track and influence government regarding fuel efficiency regulations and alternative fuels, it might be to Andy's advantage to know of alternative ways of reducing greenhouse gas emissions to suggest in place of automobile regulations. But Andy and the other automobile specialists we interviewed, like the coal specialist discussed in the next chapter, generally were focused on their own industries.

The most comprehensive set of volunteered proposals came from a congressional staff member who had written a bill on global warming to be introduced by the member of congress he worked for. He begins his answer not with specific policies but with a general strategy for choosing policies (actually part rhetoric, part strategy).

> *What do you think ought to be done?* The first and foremost thing we have articulated, myself and [name of member of congress], has been: Do those things that tie in with solving other problems. That's the best way of buying insurance, that if you turn out to be wrong, you've done these things for several other reasons, whether you're solving acid rain, urban smog, alleviating poverty, creating a larger sustainable agriculture and forestry practices, preventing species extinction or . . . All those are good reasons to do things like energy efficiency.—Mark (legislative aide)

He then laid out multiple benefits that energy efficiency would have. Having thus introduced and justified his first policy area, energy efficiency, he begins not with a list of policies but by anticipating the reactions to his (subsequent) proposals.

> I find a dismal lack of understanding of energy efficiency in that it was rhetorically supported by everyone, but it is then quickly written off as "Well yes, everybody will do that." With no appreciation for all the barriers, the fact that we subsidize heavily fossil and nuclear resources, which skewed energy efficiency.—Mark (legislative aide)

Mark then cites quantitative results from several studies indicating problems that will occur if efficiency is not improved. At that point he was ready to answer our specific question with a list of proposals, which we give a sample of below. The following excerpt is less than a third of his answer to the question "*What do you think ought to be done?*" His full answer included thirty-nine specific numerical results from scientific studies, all cited from memory.

We have been pushing for a whole range of efficiency measures. For automobile efficiency standards, for appliance standards, for least-cost utility planning. Incentive rate reform [to give utilities higher profits for energy efficiency than for new supply] . . . We're now trying to get the World Bank and the multidevelopment banks to adopt least-cost utility planning and recognize that there is a way to make it profitable to save energy. And we also have been trying to get the federal government off the dime, because we realize that the federal government can procure so many of these technologies, but it is not. . . . A vigorous R&D effort [research and development on efficiency and renewable energy] . . . Secondly, we think there is already some cost effective renewables. . . . Tree planting . . . the U.S. continues to only plant one tree for every four to five that are cut down in the urban environments. We are basically deforesting our communities for the last fifty years. . . . In the wake of that [urban] deforestation, we're now paying several billion dollars a year for air conditioning, for the ecosystems services provided by those trees previously. . . . passive solar . . . Other renewables, photovoltaics . . . is going to emerge great guns within the next ten years. It's already very appropriate in a lot of isolated cases. And, I could go on and on, [but] I don't know if you want me to. *If you could just stay at the top level, like tree planting, photovoltaics.* . . . Well we believe in a diversified portfolio. [For example,] we can see problems with geothermal. We like it. But if you go into a rainforest, or an ancient forest area, we don't like it. We like some hydro, but if you're going to build a big dam and flood rivers or block rivers, we don't like it. . . . We don't like corn-based ethanol, we do like wood-based ethanol fuels, which can probably be available cost effectively in 1997. Population stabilization is another big one. We estimate that that's one of the lowest [-cost] ways to suppress the carbon [emissions], is to provide universal voluntary family-planning services. [For] ten billion dollars a year . . . voluntary family-planning services would, according to the demographers, keep world population under ten billion, instead of the current business-as-usual which will put us up to fourteen billion.—Mark (legislative aide)

Even among the specialists, Mark's answer was very unusual in its detail and in the number of diverse policies suggested. The other congressional staff typically mentioned just a few policies, or mentioned one or two and debated the various sides and constituencies of them. Mark's answer is also highly structured: he provides first a strategy—to press for global warming policies that have other benefits—then his assessment of knowledge of others in the legislative process—poor—then a list of policies. His list is exclusively CO_2-based, not mentioning the other greenhouse gases. Other than that, his policy list is consistent with the bulk of policy analysis on the subject. His policies include energy efficiency, urban tree planting, purchases of cost-competitive renewables, and family-planning services. Within energy efficiency, he lists federally sponsored research, efficiency standards for new cars and appliances, reforms in utility regulation, and more energy-cost-wise federal procurement.

A few generalizations can be made about the specialists. All three of the specialists quoted above would be considered knowledgeable on global warming. Yet to the extent that the semistructured interviews probe it, their knowledge is limited by, and highly focused on, each informant's role in their own institution. The auto engineer focuses on fuel efficiency standards and the problems they cause his company. He also mentions alternative automobile fuels, and is proud of his company's actions to eliminate use of CFCs (not mentioning that this transition is now required by law). The advocate and the legislative aide both focused on possible new laws, and on CO_2, neglecting other gases. The legislative aide is unusual, even among specialists, in covering the topic so comprehensively. But even he narrows his focus to CO_2-related policies. In the case studies of the specialists, covered in the next chapter, we see a similar limitation of knowledge to fit institutional prejudices.

There appears to be an interesting regularity here. The specialists are ostensibly working on the same policy problem from rather different knowledge bases. Those from different institutions are seeing different parts of the picture. Why? One of our interview questions, which asked where the informant got the information they gave us on the greenhouse effect, showed that the specialists did have different sources. The advocate got his information from publications of the four environmental organizations to which he belongs and some scientific journals; the auto engineer followed internal company publications and the clipping service run by his office; and the especially knowledgeable legislative aide read research publications on global warming. All three also report getting information on global warming from the news media. With the exception of the news media, they are living in different worlds of information sources. Although semistructured interviews are not a precise instrument for quantitatively comparing knowledge, we judge the differences among them in what they know, in which subparts of the problem they are familiar with, to be much larger than their differences in opinion. The auto engineer knows little about the effects of global warming, and the advocate and legislative aide know little about the industry's reasons for opposing auto efficiency regulations. Such background information may differ more among them than the derivative fact that one side supports and the other opposes higher federal auto efficiency regulations.

Policies Volunteered by Laypeople

Laypeople do not have access to many of the sources of information cited by the specialists above. Nor are they inclinced to seek out, study, and remember specific reports. Although they may have seen a television special or read a newspaper report describing potential policies, laypeople usually would have no particular motivation to remember the details. Therefore, they might be expected to rely on cultural models to produce answers about policy, as the quotes below show. For example, the following were given as policies to combat the greenhouse effect:

We've got to stop putting all these chemicals, smokes and whatever into the air. I guess industry is our biggest problem.—James (farmer, custodian)

Try to cut down emissions of industries, ban aerosol cans . . .—Tom (college student)

Stricter rules regarding environmental pollution by individuals as well as corporations. Saving the forests at all costs. The Amazonian rain forest is essential to our existence on this planet.—Jenny (social studies teacher)

Table 6.1 lists all the volunteered suggestions given by the seventeen lay informants who were asked what to do about global warming (before the briefing). In this set of policy suggestions, we see clearly the outcomes from cultural models described in chapter 4. Reducing or controlling pollution is most popular because people are applying the general pollution model to greenhouse gas emissions. However, there are no pollution control devices that can reduce CO_2. The lay use of an ozone depletion model to understand global warming similarly reinforces the second most frequent suggestion, to ban spray cans (from which CFCs are already banned). Most policy analysts, and all three specialists cited

Table 6.1
Volunteered policies on global warming from our semistructured interviews

Volunteered policy	Number mentioning
Reduce pollution	9
Cut aerosols/spray cans	4
Solar/renewable energy	2
Save forests	1
Nuclear power	1
Ban styrofoam cups	1

previously in this chapter, begin with energy policy. Yet, of the seventeen lay people who were asked what the United States should do, only two mentioned energy policies at all (specifically, Paige mentioned "solar panels" and Doug mentioned both renewable energy and, with reservations, nuclear power). Not one of these lay informants volunteered energy efficiency, the policy area most cited by analysts.

Our survey included a question that directly contrasted the pollution model against emissions from fossil fuels. Excepting the environmentalists, about two-thirds of those surveyed believed pollution reduction to be *more effective* than reducing energy use:

116 Reducing pollution is a more effective way to prevent global climate change than energy conservation.

Earth First!	Sierra Club	Public	Dry cleaners	Sawmill workers
21	36	66	59	67

We infer that fewer of the environmentalists agree with this faulty statement because they have gotten more information on, and pay more attention to, the subject. If additional information is what explains this difference, further educational efforts may be sufficient to reduce the public's faulty use of the pollution model.

Our findings of which policies are more frequently volunteered in interviews are similar to the results from a recent study by Read et al. (1994), who asked an open-ended question about "the most effective actions" the U.S. government could take to "prevent global warming." As shown in table 6.2, the most commonly volunteered policies were "protect biomass" (presumably meaning to stop deforestation), "stop pollution," "reduce automobile emissions," and "protect ozone layer, restrict CFCs and spray cans." The sample was neither random nor national, but the respondents would if anything be expected to be more familiar with global warming than the average citizen (the study used an opportunity sample of 177 residents of Pittsburgh, 62 percent college graduates). Although Read et al. categorized their open-ended responses somewhat differently from our approach (we would have combined their three pollution-reduction suggestions, which would have made that the most common suggestion), their top three suggestions—pollution reduction, protect forests, and ban spray cans—are the same as our informants' top three policies. We consider Read et al.'s study a validation of the importance of the underlying cultural models we have proposed and their effect on lay policy choices. The suggestion of "recycling," which appeared at the top in several studies that have presented respondents with a prepared list to pick from, suggest that people are not only using cultural models like pollution for global warming but are also generally choosing policies that are "environmentally good."[2]

In short, lay informants had virtually no knowledge of the policies for global warming actually being debated. They instead refer to cultural

Table 6.2
Volunteered policies on global warming from high-education sample by Read et al. (1994)

Volunteered policy	Percentage mentioning
Protect biomass	34
Stop or limit pollution	31
Reduce automobile emissions	30
Protect ozone layer, restrict CFCs and aerosol cans	28
Reduce industrial emissions	19
Recycle	14
Facilitate public transport	13
Facilitate alternative energy	13
Nonspecific environmental legislation	11
Increase public awareness/education	10

models, such as the pollution model, or made guesses about global warming policy based on attitudes about what is environmentally good.

Postbriefing Policy Discussion

After asking our informants to volunteer their own solutions to the problem of global warming, our interview presents them with a set of five policies to evaluate. Our survey also covers many policy proposals. In both the interviews and survey, general policy strategies were included as well as specific proposals for government action. We discuss here only those for which we have sufficiently complete information to add something new to research literature.[3]

Adaptation rather than Prevention

The first choice that must be made about global warming is a strategic one: whether to try to prevent it or just adapt to the changes.[4] After our briefing on global warming, we asked our informants to evaluate a proposal for adaptation without prevention. This proposal was derived from published analyses. For example, Schelling has argued that we could adapt to global warming by expanding irrigation, building dikes, and developing new seed strains, and that future generations may be better off if we invest in other productive technology and capital, rather than spending the same money on prevention (Schelling 1990a; Passel 1989). For Manne and Richels (1989), "it is unclear whether it would be justified to incur these costs [of prevention]." These analyses apply the standard economic method of discounting future costs, which makes the future cost of adaptation appear smaller than today's cost of prevention (as explained in chapter 5). A related argument is that society in the future will be better able to afford the costs of adaptation because our grandchildren's society will be wealthier than today's, just as we are wealthier than our grandparents (Schelling 1990b). Attempting to express these ideas briefly in lay language, our interview question was:

We cannot be 100 percent certain that the climate will change. Why spend money if we may not even need to? Anyway, some effects, like warmer winters, may be beneficial. We could wait and see what actually happens, then react to that. For example, if the Midwest becomes hotter and dryer, farmers could switch to crops that require less moisture, move north, or go into other busi-

nesses. Or, if sea level rises, populated areas could be diked, as has been done in Holland. What would you think about that?

In our twenty semistructured interviews with laypeople, only two informants (John and Frank) think adaptation without prevention is a reasonable strategy. Many of the remaining eighteen informants who express negative reactions do so with vehemence. The above question tries to frame adaptation as a deliberate strategy to deal with uncertainty, mentioning that some effects could be beneficial. Nevertheless, many informants perceive adaptation as postponing or avoiding decision making.

Well, that's more or less an avoidance proposal, and apparently, we have been avoiding—me included—have just avoided this up to now, and I would rather not take a chance. If there is [another] proposal that could maybe prevent this from happening, I'd rather go with that.—Amanda (housewife)

This view persists even when the interviewer challenges an initially negative reaction.

No. [vigorous shaking of head] Because it's just supporting "Let's wait and see." Let's not do anything and wait 'till the last minute and hope that we can solve it all in an hour or two. . . . *Suppose I could convince you that we really could do something at the last minute . . . throw up a bunch of dikes to take care of the sea level and shift those farmers to other crops. . . . Would you have any other objection?* I think we're clever and we can come up with quick solutions in times of desperate need, but I'm just not a risk taker. . . . We don't know what might happen, so why not be ultraconservative up front, rather than less prepared at the end. It's too critical.—Jenny (social studies teacher)

Taking a different tack, Paige makes an analogy with military preparedness to refute our adaptation argument, "*Why spend money if we may not even need to?*"

This is not a good proposal. You shouldn't wait and see what happens. No. I mean, they're putting money into arms for wars that may not happen, and cutting school lunches to put money into bombs. This [adaptation proposal] is not good.—Paige (manufacturing worker)

The discussion of the lay value of descendants suggested that discounting long-term environmental damage violates a cultural value. Here, laypeople reject an adaptation policy for different reasons—that prevention seems a good insurance policy, similar to other social investments made to reduce the damage of uncertain future risks.

The survey results corroborate these findings from the semistructured interview. Most of the survey respondents agree that some adaptation to climate change will be inevitable.

2 We are not going to be able to totally prevent global climate change, so adaptation has to be part of the solution.

Earth First!	Sierra Club	Public	Dry cleaners	Sawmill workers
68	78	67	73	74

Nevertheless, they strongly prefer a preventive strategy, applying a cultural model something like that invoked by the traditional saying "a stitch in time saves nine."

31 Preventing global climate change is better than waiting to see what happens. It's more costly to fix problems than it is to prevent them in the first place.

Earth First!	Sierra Club	Public	Dry cleaners	Sawmill workers
93	100	97	96	85

They also reject relying on future technologies to repair environmental damage.

132 We shouldn't be too worried about environmental damage. Technology is developing so fast that, in the future, people will be able to repair most of the environmental damage that has been done.

Earth First!	Sierra Club	Public	Dry cleaners	Sawmill workers
0	4	10	13	15

In the semistructured interviews, and in the following survey statement, we tried to explain the logic behind discounting and adaptation in understandable terms. To the extent we succeeded in explaining the logic, respondents reject the economic rationale of discounting future environmental costs.

28 We should invest in industry rather than spending money on the environment, so that our economy will grow. Our children would then be more prosperous and better able to afford the cost of fixing any environmental problems we may have caused.

Earth First!	Sierra Club	Public	Dry cleaners	Sawmill workers
0	0	25	14	15

Whatever the causes, the negative reaction to the adaptation proposal is broad (eighteen of twenty lay informants in the semistructured interviews; 75 to 97 percent in survey questions 28 and 31). The vehemence of some reactions shows that the rejection of this strategy is often strongly felt. Furthermore, despite our portrayal of it in the briefing as a coherent, planned strategy, comments in the semistructured interview show that adaptation was perceived as avoidance of decision making by politicians. For proponents of adaptation in place of prevention, these results suggest that publicly advocating this strategy would be politically difficult. For opponents of adaptation, an effective public campaign might only require communicating to the public what adaptation is, and what prevention policies are being passed up.

Wait for More Evidence

A more moderate version of the "adapt rather than prevent" policy is to defer a decision, leaving open the possibilities of prevention (soon), adaptation (later), a combination, or other options. Supporters of this strategy argue that, because the climatic predictions are uncertain, we should wait for better data before committing resources. In fact, this was essentially the position of the Bush administration, which was in office at the time of the interviews. Although we did not ask explicitly about waiting for more evidence as a separate policy in the semistructured interviews, some informants interpreted our adaptation proposal that way. Consequently, in our survey we included a series of explicit statements dealing with parts of the "wait for evidence" argument.

Except for the environmentalists, about half of our survey respondents recognize a diversity of scientific opinion and agree that being "skeptical" is appropriate.

88 We should be skeptical about scientists' predictions of climate change. You can get five different estimates from five different scientists.

Earth First!	Sierra Club	Public	Dry cleaners	Sawmill workers
30	26	50	53	63

Nevertheless, in responses to the following series of related survey statements, at least 70 percent of all groups favored going ahead with actions, even if the evidence is not yet definitive. First, substantial majorities felt there was "enough research" to at least "start" action.

18 The scientists have already done enough research on global climate change to start doing something.

Earth First!	Sierra Club	Public	Dry cleaners	Sawmill workers
94	78	87	70	78

142 We should have started dealing with the problem of global climate change years ago.

Earth First!	Sierra Club	Public	Dry cleaners	Sawmill workers
100	93	87	93	81

Similar percentages disagreed with two variants of the converse statement.

90 Scientists are just speculating about global climate change. We shouldn't take actions until they have proof.

Earth First!	Sierra Club	Public	Dry cleaners	Sawmill workers
0	0	21	20	22

38 So far, global climate change seems to be only a theoretical possibility. We shouldn't make major changes in our economy and in our way of life for a theoretical risk.

Earth First!	Sierra Club	Public	Dry cleaners	Sawmill workers
0	7	24	13	48

Significant to interpreting the above three statements, large majorities feel that citizens can themselves make judgments, based on science reporting available to them.

134 The average person can make a decision on whether they think we should do something about global climate change, even if they don't understand all the science.

Earth First!	Sierra Club	Public	Dry cleaners	Sawmill workers
94	96	93	73	78

One argument for waiting for more evidence is that scientists differ on whether there is a clear indication of changing temperature yet, and that such a signal would be unambiguous evidence for global warming. Some climatologists have speculated that political support for tough global warming policies will require a clear temperature signal. Most of our respondents reject this recommendation.

130 We shouldn't take action on global climate change based just on computer models and scientists' predictions. We should wait until we can measure an actual temperature change and know its real.

Earth First!	Sierra Club	Public	Dry cleaners	Sawmill workers
0	11	27	20	30

These survey statements address complex and critical policy questions regarding global warming: whether or not, and how long, society should wait for more definitive evidence, and whether prevention should proceed on the basis of models rather than waiting for definitive temperature measurements. These questions are not explicitly on our semistructured interview protocol, so we do not have sufficient interview data to understand the cultural models and values behind our survey results. Nevertheless, some reasons are suggested by the strong preference for prevention stated in the prior section, and the trust in scientists described at the end of this chapter.

It is interesting to compare these preferences with those of scientists. A poll of scientists studying climate in 1990 reached a similar finding: Only 65 percent expected a 2°C rise at better than fifty-fifty probability, yet 90 percent favored taking immediate steps to reduce CO_2 emissions (Holden 1990). Another poll of atmospheric scientists by Stewart et al. (1992) found similarly that 62 percent agreed there was "little doubt . . . of global temperature increase," yet only 19 percent concurred on delaying preventive action because of these uncertainties. If we can assume that most of the climate or atmospheric scientists know that the temperature signal to date is ambiguous, our survey respondents give logically identical results to these polls of atmospheric scientists: while a slight majority expect climate change to occur, much larger percentages want society to start doing something about it. Both scientists and laypeople say they want society to begin doing something now on the basis of existing computer models, without waiting either for a definitive temperature verification or more refinements of the computer models. We state this as an interesting point of congruence between laypeople and scientists. Our point is not to say that polling illuminates which policy strategy is the correct one—in fact, as a policy recommendation it is meaningless without quantitative values for *how much* prevention should be pursued now. But in an area involving complex interplay of the quality of scientific evidence, models versus measurements, and values, it is remarkable to see so much similarity, at such a high level of detail, between scientists and the public.

Lifestyle Change

A third general strategy question regarding global warming is whether changes in lifestyle would be required to achieve sufficient reductions in greenhouse gas emissions. In the semistructured interview, we raised this question by opposing changes in "the way we live" to "better technology," as alternative solutions to global warming.

Some people say that the best way to solve this type of environmental problem is with better technology; others say we need to change the way we live. What would be your opinion?

This question generated a range of responses. One set was that Americans are spoiled, cannot change, or should not have to.

It would probably be better technology, because I don't think anybody's going to change. They're just too spoiled.—Cindy (housewife)

We can't change the way we live. Nobody's going to make anybody. *We could drive cars less or something like that.* We could what? *Drive cars less.* Less? *Yeah, use some other form of transportation.* No way, not for me.—John (retired factory foreman)

The argument that lifestyle was unchangeable could also lead to concluding that technology is the only possible solution.

I don't think the way we live is going to change appreciably. Maybe some, maybe if we use the technology that's being developed, we can change the way we live to a degree as far as what we would use for our fuels and that sort of thing. But I think by and large, it has to be left up to technology. . . . I think technology is the answer. There's just so many people, and I don't think people live that much differently today than they did before this problem [was] here. We haven't changed; technology hasn't kept up with, with the needs of the environment.—Peter (logging contractor)

Another set saw technology as providing only a minor component, or saw technical changes as already exhausted, or saw lifestyle change as a more "basic" change. Some also made the related point that there are a lot of opportunities for saving energy via lifestyle change.

Change the way we live more than the technology. Technology can help out, but a lifestyle change is important and an attitude change too. Technology can help if we were gonna expand on solar . . . scrubbers and energy-type things. But the basic change has to come from . . . more of an attitude, lifestyle change than technological. . . . Conserving and recycling and all that.—Marie (health inspector, environmentalist)

We're at the point now where we've squeezed out just about everything we can out of these large sources—at least with respect to air [pollution]—and we're at the point now where we have to start looking at ways that would have the effect of changing people's lifestyles. And that's where the politicians, they get a little edgy. [interviewer laughs] Quite frankly.—Gerard (legislative council)

Some environmentalists saw the typical American lifestyle as being excessive and as being determined by social forces rather than individual choices.

People are not criminal. They buy a fancy gas guzzler, but . . . it's just that society is set up the way it is so that people are encouraged to buy fancy cars, and people are encouraged to drive places instead of take trains. I think our society is just set up to encourage a lot of waste and excess consumption which is just totally unnecessary.—Marge (lawyer, environmentalist)

I guess I have a vested interest in staying enough in the mainstream to be able to go out . . . informing and educating. [However,] it's just about impossible [living] your average lifestyle to be a really environmentally sound person.—Margaret (activist, environmentalist)

In contrast, some of those who opposed environmental regulations saw environmental goals as disguising an agenda for social change. That is, they thought environmental reasons were being given as an excuse for social changes actually desired for other reasons. A mild form of this accusation is expressed below, in this case referring to mandated changes to automobiles for emissions and fuel economy.

In my mind some of the people who are making the proposals have a different . . . agenda. . . They don't mind changing lifestyles, or the public mind's changing their lifestyle. *Right.* And the message coming across doesn't say: "Hey, guys, you're not going to live in your three-thousand-square-foot air conditioned house, or drive your six-passenger car." . . . *There's a lack of honestly expressing this to the public.* Right; it's as if there's a fear that the public will say, "no thanks."—Steve (auto fuel efficiency planner)

In our survey, we presented a series of statements to investigate further a general strategy of favoring changes in lifestyle for environmental reasons. On the first two points below, Americans seem to be of mixed opinion, as indicated by seemingly contradictory responses as wording or context is changed. For example, a majority thought Americans are too "spoiled" to change lifestyle; roughly the same numbers saw Americans as being willing to ration to save the environment.

67 Americans are too spoiled to change their lifestyle.

Earth First!	Sierra Club	Public	Dry cleaners	Sawmill workers
74	74	63	67	56

22 In World War II, people gladly rationed for the war effort. People would willingly do it again to save the environment if the need were great enough.

Earth First!	Sierra Club	Public	Dry cleaners	Sawmill workers
77	67	57	83	93

In the other example of a seeming contradiction, a majority of all groups but Earth First! agreed both that we will, and will not, have to make reductions in how we live.

37 We don't have to reduce our standard of living to solve global climate change or other environmental problems.

Earth First!	Sierra Club	Public	Dry cleaners	Sawmill workers
23	59	60	63	67

147 Americans are going to have to drastically reduce their level of consumption over the next few years.

Earth First!	Sierra Club	Public	Dry cleaners	Sawmill workers
90	74	87	69	67

The seeming contradiction between statements 37 and 147 may be resolved by postulating that respondents believe that reducing our "consumption" can be done without reducing our "standard of living."

A series of other statements explored altering lifestyles or consumption patterns as an environmental strategy. For example, cultural changes are seen by large majorities of each group as being environmentally beneficial.

141 We should return to more traditional values and a less materialistic way of life to help the environment.

Earth First!	Sierra Club	Public	Dry cleaners	Sawmill workers
100	85	80	79	70

Opinion is divided as to whether lifestyle change completely eliminates the need for technical improvements, with a slight majority (of all groups but Earth First!) thinking that new technology will still be necessary.

71 If we could get people to change their lifestyle, we wouldn't need new technology to prevent global climate change.

Earth First!	Sierra Club	Public	Dry cleaners	Sawmill workers
61	42	38	43	41

One of the more surprising lifestyle findings is that majorities of all groups, and three-quarters of our public sample, apparently are willing to "force" change in lifestyles for the environment, a view that seems to override American values of liberty and freedom.

103 You shouldn't force people to change their lifestyle for the sake of the environment.

Earth First!	Sierra Club	Public	Dry cleaners	Sawmill workers
0	0	27	13	41

A few in the semistructured interviews mentioned radical lifestyle changes, such as going "back to colonial times," as a way to reduce energy consumption and deal with global warming. The survey shows that very few see the colonial past as a desirable option.

15 We're advancing so fast and are so out of control that we should just shut down and go back to the way it was in colonial times.

Earth First!	Sierra Club	Public	Dry cleaners	Sawmill workers
42	11	14	10	22

Even though such radical statements are often attributed to Earth First! members, we note that the majority of them disagree with it. One reason a return to the past is rejected is that few people in any group believe even a major environmental problem like global warming would require such drastic measures.

136 We really cannot prevent global climate change, because we'd have to shut down modern society.

Earth First!	Sierra Club	Public	Dry cleaners	Sawmill workers
13	15	23	17	26

However, despite the strong support found for lifestyle changes, our respondents are sympathetic to basic "necessities:"

138 There are certain basic necessities, such as heat, that people cannot give up no matter what the environmental costs are.

Earth First!	Sierra Club	Public	Dry cleaners	Sawmill workers
32	63	76	70	67

Some of the quotations from our interviews, and the complex pattern of survey responses, suggest that there is much more to the question of lifestyle change than we have captured here. American opinion is divided as to the potential contribution of changing technology versus "the way we live," perhaps in part because our questions never precisely define the latter. Majorities believe that consumption changes, "traditional values," or "nonmaterialism" would have environmental benefits, that reductions in consumption will be necessary, but that we should not (and need not) "shut down" modern society.

We find these responses intriguing in conjunction with other recent studies. For example, De Young (1991) has found that Americans view environmentally conscious, lower-consumption lifestyles as being more satisfying, a finding in opposition to the stereotype of "giving up" for the environment. Several cross-national and cross-cultural studies suggest that this could be factually correct, that lower-energy or lower-consumption societies may be more satisfying to live in (Nader and Beckerman 1978; Sahlins 1972; Johnson 1978). And some changes in individual consumption patterns—though arguably not major ones—have already been made voluntarily in order to achieve collective environmental benefits (Kempton 1993).

These studies, along with findings from our interviews and survey, suggest that environmental policies intended to stimulate major lifestyle changes deserve more serious consideration than they have received to date. Most mainstream environmental activists, and virtually all analysts, have avoided suggesting any change in level of consumption. In this area, the public at least claims they are willing to do what the advocates will not promote and the analysts have not studied, with substantial majorities of even the least environmentally inclined groups supporting "a less materialistic way of life to help the environment." It remains to be seen whether we can believe these verbal claims and by what means they might be translated into action.

Energy Use and Efficiency

When news media coverage of global warming mentions solutions, it usually mentions energy efficiency as the best, or a primary, strategy to reduce the use of fossil fuels and hence to combat global warming. This has been true since the early in-depth news coverage, as global warming first began to be known widely (e.g., *Newsweek* 1988; Koppel 1988). Perhaps the most surprising policy-related finding from the semistructured interviews is the public's lack of connection of global warming with energy. Burning fossil fuel is not seen as a cause, and efficiency is not seen as a solution. This lack of connection was evident in our semistructured interviews, when we asked for ways to avoid global warming, only two of our twenty lay informants cited any type of energy solutions (table 6.1). Similarly, only 13 percent in the study by Read et al. (1994) mentioned energy (table 6.2). Although our survey and previous polls show that people have positive attitudes toward energy conservation, they do not really understand how conservation works or even the relationship of energy to global warming.

Our briefing summarizes current scientific knowledge on global warming. We show a pie chart of causes, with fossil fuels the biggest part, while reading "*The biggest part, over half, is burning coal, oil, and natural gas for energy.*" Informants' responses during and after our briefing suggest that we were able to convey the connection between energy and global warming to most informants, but that this was news for most of them.

People are burning that much coal? Why? *Well, a lot of the electricity we get is from burning coal. . . .* Oh, oh, . . . I was still thinking, water power. . . . I wasn't aware of it. All I knew was, you turn the light on, and that's it, you know, I don't know where it came from . . .—Paige (manufacturing worker)

What was new to you? . . . Energy use, I guess I should have known. Maybe not that big of a percentage. Others I guess I was aware of.—Tara (sales manager)

[interrupting discussion of fuels as greenhouse gas sources] Is it? That's what's causing the greenhouse? *It's more than half the problem, yeah.* Burning gas? So that's like our heating and all that? *Right. Also gasoline, coal in power plants, oil in power plants.* Well, there's nothing that the government will do about that. Nobody's going to stop working, right?—Cindy (housewife)

This lack of connection of global warming to energy has been missed in national surveys, partly due to the confounding effects of the cultural model of pollution. For example, in a previous survey by RSM, respondents were asked: "Some people say that global warming and the greenhouse effect are caused by the destruction of the great rain forests of the earth. Others believe it is caused primarily by fossil fuel emissions from power plants and exhaust gases from cars and trucks. Which of these two views comes closest to your own opinion?" In response, 60 percent said fossil fuels, and 27 percent said loss of forests. From these percentages, the report's overview summarizes that "American voters continue to be sensitive to the role of fossil fuels in the generation of atmospheric pollutants and their contribution to global warming." (RSM 1989). We contend that Americans connect global warming to pollution but not to energy consumption. Given the two-way choice in the RSM survey question, we suspect that respondents invoked their pollution model and chose "emissions . . . and exhaust gases." Our survey question 116, cited earlier in this chapter, distinguishes better than RSM's between controlling pollution and reducing consumption. To that question, majorities of all but environmentalists agreed that "reducing pollution is a more effective way to prevent global climate change than energy conservation."

The quotations above, like that from Cindy ("Nobody's going to stop working"), illustrate a second problem in public discussion of energy efficiency as a solution to global warming. People do not understand what energy efficiency is because they have had very little direct contact with it. A majority of energy analysts, though not all, use "energy conservation" to refer primarily to efficiency, that is, to getting the same service from less energy, by better technical devices or better management. Laypeople tend to interpret energy conservation as decreasing energy services, that is, curtailment or sacrifice. This problem was identified a decade ago (Stern and Gardner 1981; Kempton et al. 1982), and our survey makes clear it is still predominant, even among environmentalists.

115 Energy conservation means doing without some things that give us comfort and enjoyment.

Earth First!	Sierra Club	Public	Dry cleaners	Sawmill workers
77	63	73	63	78

How do lay people imagine "doing without some things"? An austere past is the best cultural model available to conceptualize a low-energy future.

It's hard to believe that our way of life is affecting the environment as drastically as it is. . . . I'm not sure how to change my lifestyle . . . without going all the way back to colonial times. Or, I don't know, a different way to produce energy for us, which I'm sure they're developing.—Amanda (housewife)

The above quote about "a different way to produce energy" is one of several examples of solar or alternative energy being mentioned after the briefing, but not energy efficiency. This suggests that, after being briefed on CO_2 as a cause of global warming, people can more easily conceptualize alternative energy sources replacing carboniferous sources than they can conceptualize energy efficiency. Energy specialists know that it is usually cheaper to eliminate need for a kilowatt hour via efficiency than to produce a clean kilowatt hour via new renewable energy sources, and efficiency is often cheaper even than today's polluting energy sources. That idea clearly was foreign to our informants, even after the briefing, so it does not seem to have gotten out to the broader society yet.

The one layperson who clearly understood efficiency was Nick, owner and operator of a light manufacturing facility making custom wooden furniture. Nick's volunteered policies were not effective ones because he confused global warming with ozone depletion. However, in the post-briefing discussion of our suggested policies, he had a good understanding of technical energy-efficiency improvements. For example, he generalized our proposal of automobile standards for increased MPG in just the way an analyst would, summarizing it as "basically solving the problem [of global warming] through higher energy efficiency." He views regulations as helping to stimulate technical innovations in efficiency. Whether or not one agrees with his view of regulations, Nick has a good understanding of the potential of technical efficiency and one of the policy contexts in which it occurs. Speaking of the probable effect of our proposed policy of automobile MPG regulations, he says:

> [Regulations] in balance with economics, that's the right way to go about it. . . . With all the pissing and moaning that Detroit has done and everybody else, [nevertheless,] here we are and it's merely ten years later and the mileage of all the cars on the road is double what it was ten years ago. So it does work. And, yeah, they cost more, but the Japanese still do it for less.—Nick (small manufacturing plant owner)

Why does this one person, out of all our laypeople, summarize the automobile regulations as "energy efficiency" and see it as a policy solution to global warming? We believe it is because he has had recent experience with a dramatic efficiency change in a carefully measured part of his business: application of lacquer to furniture. Nick's shop had been using traditional spray equipment, which transferred 30 percent of the mixture sprayed onto the surface, the rest going into the air. Also, he had originally been using standard lacquer, which is 20 percent solids and 80 percent thinners as a vehicle. As he summarized, "So you are putting 7 percent on your board and the rest blows up [into] the air."[5] Due to very strict air quality regulations in Southern California, high-efficiency spray guns were developed that put less thinner into the air. "Nobody was making one of these guns until all the body shops in California all of a sudden can't spray cars. And then all of a sudden there's a market. . . . Then somebody dumps the R&D [research and development costs] to go after the market." He saw these spray guns advertised in trade magazines and got one, apparently both because he wanted to "do the right thing," as he put it, and because he thought such regulations would later be enacted in his state.

Nick had measured the differences between his old and new paint guns, verifying the claimed advantages.

> So now we are using high-efficiency guns, which can transfer as high as 90 percent. From 30 percent. The quality's not quite as good, but for what [we use it for], simple finishes, it's close enough. And we're using new materials that have 40 percent solids in them. So now we put 90 percent of 40 percent on, so we're doing five times what we used to do before.—Nick (small manufacturing plant owner)

He also notes that demand for these high-efficiency guns, along with market competition, brought down the prices: "when they first came out they were eight hundred bucks and now they're two-fifty. If you're doing enough spraying to make it an issue, you've long since paid for it with the savings in materials."

Nick's spray gun provides him with a concrete example of how much technical efficiency improvements can save. It also illustrates to him that government regulations, in combination with market competition, can stimulate technical innovations that both "do the right thing" environmentally and reduce production costs. Nick applies this example to several environmental issues and proposed policies we discuss. In particular, it makes his discussion of energy efficiency dramatically more complete than that of other lay informants.

Notice that we are discussing whether laypeople understand energy efficiency, not whether they are in favor of it. Our survey questions 6 and 33 (see appendix C) show near-universal support for "efficient" use of resources and "energy conservation." This can also be seen in national polls that have asked respondents to compare energy options (Farhar 1993). When the polls include solar, various conventional energy sources, and "energy efficiency" or "energy conservation," solar consistently ranks as most preferred and efficiency or conservation comes in second. Our point is not that consumers oppose energy efficiency and conservation—indeed, they seem to have a good feeling about it—but rather that they do not really know what it is or how to think about it.

In sum, analysts may favor energy efficiency as the fastest and most cost-effective path to rapid CO_2 reductions in the United States, possibly as a step toward a renewable energy supply system. However, anyone publicly advocating this solution will need to clearly spell out what energy efficiency means and its connection with global warming.

Automobile Efficiency Standards

Despite informants not seeing the connection between energy conservation and global warming, they do support many energy-efficiency proposals. The one energy-efficiency proposal we described after the briefing was broadly supported.

Another idea would be for government regulations requiring higher energy efficiency to reduce the amount of fuel required by new products. The regulations might apply to new cars, houses, and appliances. Take cars, for example. Suppose new cars were required to get at least 55 MPG, thus cutting the amount of gas burned by half from today's new cars. However, let's say that this would mean that the next new car you bought would be a little smaller, and it would not be a high-performance car. What would you think about that?

The wording here was carefully designed. We took an efficiency level above that normally discussed (the current U.S. new car fleet average is 27 MPG, by regulation), but we were careful to emphasize that this was not "free" to the consumer, that there would be performance penalties. Typical reactions were:

It [smaller, lower-performance cars] wouldn't bother me a bit; I'm a small-car person. For me a car is four wheels on a frame to get me from point A to point B. If you look at the general population, you'll see more smaller cars than large cars. If you want to impress someone, you rent a limo.—Susan (hospital administrator)

It wouldn't bother me none. . . . *The loss of performance wouldn't bother you?* I think that would have to be addressed. There would have to be some sort of adjustment made to the size of a family. . . . When you've got four or five kids in the family you can't stick everybody in a Volkswagen.—George (coal mine truck driver)

The support we found is consistent with national polling. For example, 83 percent of a national sample favored a 45 MPG fuel standard when told that cars would cost $500 more but the added cost would be recovered in gasoline savings over four years (RSM 1989).

Consistent with the RSM poll, we found a few who opposed automobile efficiency requirements. In our semistructured interviews only two of the twenty lay informants, Ronald and John, strongly opposed this policy. John was most picturesque.

Well, I got a Cadillac. Fleetwood. I wouldn't feel safe in a little coffin. Matter of fact you look at these people that are killed. They're all little cars. Every one.—John (retired factory foreman)

Whereas John opposed auto fuel standards for pure self-interest, the other opponent, Ronald, added a more principled opposition. He called the proposal "terrible," explaining:

Because it seriously impacts on my convenience. And it seems to me it would be like bailing out the ocean with a thimble. . . . If we indeed have conclusive evidence that there is this greenhouse danger . . . then we have to go at it seriously, [not] just jiggling things like car mileages.—Ronald (resort proprietor)

Behind Ronald's conclusion are three assumptions: (1) high MPG automobiles would be smaller and less "convenient," (2) the amount of energy saved by fuel efficiency standards would be small in relation to national consumption, and (3) there are other, larger, and "more serious" options to save more energy than auto efficiency. Comparing these three points to recent technical literature on automobile efficiency (e.g., National Research Council 1992; DeCicco and Ross 1993; Plotkin 1993), the first may be true, depending on the design and how one defines convenience. The second and third are incorrect: transportation uses roughly a third of national energy and two-thirds of petroleum, with private automobiles consuming the bulk of transportation energy.

Like the national poll by RSM cited previously, our survey responses illustrate that Ronald's position is a minority one. The survey had two questions specifically asking about auto efficiency regulations, each testing a line of argument against regulations. The first one made the case for shifting the responsibility to individual drivers.

129 We should not force the auto companies to make cars with higher gas mileage. Instead we should discourage people's excessive use of their cars.

Earth First!	Sierra Club	Public	Dry cleaners	Sawmill workers
32	22	33	37	30

The second question was worded in a way advocating minimal regulation, but responses showed little acceptance of the ideological argument that efficiency regulations are too intrusive.

82 The problem I have with requiring fuel efficient cars is that I would resent the infringement of my personal liberty to be told I couldn't drive a particular kind of car.

Earth First!	Sierra Club	Public	Dry cleaners	Sawmill workers
0	18	27	30	41

There is a regular scaling of support across our five groups, but in no group does a majority accept this argument.

An entirely separate factor may add to the broad support for automobile efficiency. In the semistructured interviews, some informants volunteered a conspiracy theory for why more efficiency was not practiced.

I hear rumors that they could easily design a much more fuel efficient car, but it's not always done because [of] the money and politics and that kind of thing. . . . Keep the oil companies going and, and that kind of thing. I don't have any facts . . . but I think I could believe . . . that the reason we don't have more fuel efficient cars is because the car companies want to stay in business selling more cars, and the oil companies want to sell more gas, and they'd like, they like us to be dependent on fossil fuels.—Marie (health inspector, environmentalist)

We feel that much more automobile fuel efficiency is possible, and that U.S. progress toward fuel efficiency has been driven by government, with industry consistently resisting. However, having reviewed the technical issues and discussed this issue with both independent scientists and auto employees concerned with fuel economy, we consider a conspiracy between automakers and oil companies highly unlikely. Nevertheless, we added it to the closed-ended survey to see how widely this belief is shared. We were astonished to find it accepted by overwhelming majorities in all groups.

34 They could easily design a much more fuel efficient car, but the big automobile companies conspire with the big oil companies to keep that from happening.

Earth First!	Sierra Club	Public	Dry cleaners	Sawmill workers
83	82	80	90	89

Notice that this conspiracy belief, like other negative views of industry described in this chapter, is also widely accepted by the dry cleaners and sawmill workers. Those whose industries have been adversely affected by environmental legislation nevertheless believe in antienvironmental conspiracies by other industries.

This belief is known to some in the automobile companies themselves. Steve has even coined the term "magic carburetor" for this myth.

> I still get calls. I used to work in carburation. People still ask me about the magic carburetor. It gets two hundred miles per gallon. . . . These are people with graduate degrees that are convinced it's there. And, [I'm] telling them, "Nope; we'd be selling it if it existed."—Steve (auto fuel efficiency planner)

Steve's reaction to this story is one of benign amusement with his fellow human beings. But as anthropologists, when we see that a myth is persistent and widespread, we look for some structural reason for it. For example, this myth could be expressing a deep-seated mistrust of corporate profit motivations, and fear of conspiracies among the most powerful corporations, to the detriment of consumers.

This structural cause is also suggested by the fact that another informant, more knowledgeable than average about efficiency, seemed to invent a similar explanation for lack of efficient appliances.

> I've read a lot of stuff about how energy efficiency could . . . make a big difference as far as whether we need to build new power plants. . . . You know, it just seems so ridiculous that we're not pursuing that more. It just makes no sense. . . . *Why do you think we're not pursuing it?* I don't know. I assume it's industry pressure, but I just don't know. Maybe it's just 'cause, people just don't know enough yet. . . . I don't know why industry wouldn't want to build energy efficient appliances. Or maybe it's the coal companies. . . . Whoever wants to go out and build new power plants. They don't want to be told, "Sorry we don't need any new power plants," I guess.—Marge (lawyer, environmentalist)

Note that here, as with the widespread belief in an automobile–oil industry conspiracy, there is an assumption that the product-manufacturing industry (cars or appliances) would naturally want to produce the most efficient products. So their failure to do so is explained by a conspiracy with a fuel-supply industry (gasoline or coal).

In our judgment, the reality is more complex and originates with the manufacturers themselves. Changes would require more radical design changes than replacing a carburetor. Although we cannot provide a full analysis here, we offer alternative explanations for the sake of contrast to the lay conspiracy theories. We suggest that manufacturer resistance to efficiency improvements probably derives from factors such as the expense of research and development, long product development cycles, lack of foresight by management of industries with stable product lines, separation of engineering and advanced design from government lobbying departments, and unfamiliarity of marketing departments with selling environmental benefits as product attributes.

In conclusion, cultural models relating to automotive fuel efficiency merit more analysis than our present data allow. We can conclude that there is broad support for fuel economy standards, even when traded off against higher first cost or lower performance. Myths about corporate conspiracies seem to reflect a mistrust of large corporations to pursue efficiency without government prodding, and these myths in turn lead to further support for efficiency regulations. More generally, although energy efficiency is not well understood in the abstract, automobile fuel efficiency standards are a specific policy strategy that is understood, and for which government regulations make sense to lay people.

Fossil Fuel Tax

Another specific policy proposal offered in the semistructured interviews was a 100 percent energy tax on electricity, natural gas, and gasoline. Our interviews were carried out prior to the proposal for a "Btu tax" in the early months of the Clinton administration. The proposed Btu tax, like our interview proposal, would tax all fossil fuels, but it was much smaller than the 100 percent tax we asked about. Initial press reports of public reaction to the Btu tax suggested surprisingly little resistance, perhaps due both to an acknowledged need for deficit reduction and to presidential candidate Perot's advocacy of a large gasoline tax during the presidential campaign. In time, however, an effective coalition of fuel producers and large industrial fuel consumers, allied with oil state representatives, helped whittle the proposal down until it was only a small increase in the existing federal gasoline tax. The defeat of the Btu tax proposal can be explained in terms of national politics (President Clinton did not commit to it, industry focused media campaigns on districts of undecided representatives, etc.). We deal with a different level of explanation here—why public opinion did not support an environmentally beneficial policy proposal.

After our briefing, we explained the rationale for an energy tax—that people use less when prices are higher—and then presented the specific policy.

One argument is that people and companies use less when prices are higher. Therefore, one idea is to increase the price of fuels according to how much of these harmful gases they produce. Suppose there was a 100 percent tax on gasoline. How much do you usually pay when you get gas? [*Get $ figure.*] *So a 100 percent tax would mean you would pay _____ instead of _____. What would you think about that?*

The 100 percent level was deliberately chosen to be extreme, to force informants to balance environmental benefit against significant personal expense. (A more logical tax level would be at the "social cost," that is, a tax based on estimates of the environmental, national security, and other costs of burning each fuel.[6]) We translated our proposed 100 percent tax into the informants' own increased expenses in dollars, to make the personal cost very concrete.

To separate out the issue of government's use of the revenues, after reactions to the above tax proposal we asked:

Suppose the 100 percent tax went straight into developing technologies that do not cause the greenhouse effect. Would this affect your reaction?

After the questions on gasoline, we asked the size of their home utility bills and solicited their reactions to a 100 percent tax on electricity, natural gas, and home fuel oil. The resulting comments on home fuels are sometimes used below to contrast with the reactions to a gasoline tax.

It wouldn't stop people. It wouldn't stop me. . . . I commute to work, so I have to have gas. It's not a good idea; it wouldn't accomplish its object. . . . *Suppose it were spent on research to prevent the greenhouse effect?* I don't trust the government. . . . I think it should be private industry, not the government.—Susan (hospital administrator)

No, I wouldn't be interested in that proposal, and I wouldn't use less gas. . . . That would really anger me if I were taxed on my gas. . . . Just so I can take my children to school and visit my mother. [later] Well, it doesn't have to be 100 percent? *No, it doesn't. I'm taking an extreme figure.* OK. Higher [tax] on gasoline or higher tax on utilities, that would be fine with me, if that money were to be going towards [solving the problem].—Amanda (housewife)

No, no, no, no [laugh]. . . . It doesn't matter how high gas would go, I would still buy it, just like cigarettes. . . . You have to go to work. *Suppose the 100% tax went straight into developing technologies that don't cause the greenhouse effect. Would that change your reaction at all?* [moans] That's hard, I don't think I'd like a 100 percent tax on anything. If they raised the rates a little bit and said, this money is going to prevent the greenhouse, yeah, I would go for that, but I still don't want 100 percent.—Cindy (housewife)

These reactions have a common structure: the negative response is immediately followed by a statement about inability to reduce one's own personal gasoline consumption.

In the quotations above, Amanda and Cindy reluctantly accepted an energy tax if it were not as high as 100 percent. Others accepted the proposal with few objections or preconditions.

I wouldn't be happy. [laugh] It's a lot more money, but you know, if the end result was that it was going to be beneficial, you live with it, you know.—Jenny (social studies teacher)

My fuel bills [gasoline] each month are close to a couple hundred dollars. So that would be a major cut in my belt. I wouldn't like that, but again if it could help. I don't know that they'd have to do a 100 percent increase, . . . I couldn't handle a 100 percent [gasoline] tax increase on the salary that I make now. . . . [*would it affect how much you use?*] . . . my job [is] to call on accounts, and most of the time you can't do it by phone, I mean I do some work on the phone, but most of the time I'm out on the road so, whether it costs me $200 [or] $400 a month for gas, I'm gonna have to do it, I guess.—Tara (sales manager)

Since this question was designed to highlight the negative impact on the individual, by using a 100 percent level and explicitly calculating their increase in bills, we found the reactions surprisingly mild. While some are unequivocally opposed, others accept this policy immediately, and still others would accept it if the tax revenues went to help the global warming problem or if the tax rate were lower. Acceptance may be higher due to our previous briefing on global warming problems in the interview.

For some informants, spending tax revenues on global warming made the energy tax more acceptable. For instance, Ronald vehemently objected to fuel taxes as hurting the economy.

> It threatens the economic and political power of the United States to take these extreme economic measures that are going to drain our economic life blood in the United States. And if other countries don't do this, they're going to get the upper hand politically. . . . We just don't have the economic resources to become environmental extremists or radicals. It's too costly.—Ronald (resort proprietor)

But when presented with the variant of the policy proposal of spending revenues on "technologies that do not cause the greenhouse effect," he became more receptive, saying:

> If the danger [of global warming] is real, then we've gotta take some pretty drastic actions to correct it, and it's got to be done with the aid of technology.—Ronald (resort proprietor)

Although this is a particularly dramatic reversal, Ronald is one of many who could see an energy tax in opposite ways, depending on the specifics of the proposal, what problems it was addressing, and what other policy solutions it was to be implemented in conjunction with.

The surprising finding is that not one of the twenty lay informants mentions getting a more fuel efficient car to reduce gasoline costs. This is especially surprising considering that the preceding question asks about regulations requiring a 55-MPG car, and that question explicitly says it would "[cut] the amount of gas burned by half." There might be a delay between when gasoline prices rise and when someone purchases a higher-efficiency car. Nevertheless, we are surprised that not one person responded to the tax by saying anything like, "I'm not worried about gas prices doubling because you just said our new cars will need half as much gas."

Economics provides terminology to describe this lack of connection, but does not explain it. Our informants react to the energy tax by associating it with what an economist would call short-term elasticity (e.g., driving less), never with long-term elasticity, for example, higher MPG, which is long term because it must await the next car purchase. But why did not even one lay informant volunteer the long-term elasticity solution?

One discrepancy between analyst and lay perspectives was in understanding the purpose of an energy tax. We began the tax proposal with

"*. . . people and companies use less when prices are higher.*" Several informants refuted this assertion by saying higher cost would not affect their consumption, while others seemed to ignore the assertion. For example, Cindy reacted to the gasoline tax by saying "That's a necessity, how can they tax that?" Instead of a pricing mechanism, the tax was seen as a punishment or as a way of raising money. In short, perceived inelasticity of energy consumption apparently contributes to the political opposition to energy taxes.

Even a person who had an analyst's view of market mechanisms provides a clear description of the "essentials" view. Marie saw the fuel tax as hurting the consumer, whereas regulations would not.

> There's a lot of people out there who can't afford paying their gas bill as it is, and they're freezing to death. . . . This first one [our proposed policy of auto efficiency], it's not hurting anybody's purse. I mean, it's not hurting the consumer. Whereas in the second one [fuel tax] the consumer is the one that's paying all of it. Right? *Earlier, you said you supported market solutions. . . .* People need to heat their houses and they need to have electricity. . . . My idea when I said "regulate by the market" was to increase the cost of things that are not so essential. You know, heating and gas are considered essentials. . . . toys or packaging . . . products that we don't need . . . VCRs, that kind of stuff . . . [she admits that those products would not affect a lot of energy use] . . . I mean this [energy tax] might be OK, but it just seems like it's really going to hurt. It's not going to hurt me. I'm willing. But I just think it's going to hurt a lot of people.—Marge (lawyer, environmentalist)

There is a basic distinction between essential and nonessential products, with Marge seeing energy as essential. She then tries to imagine a market system that would be closer to what she intended.

> Having products out there on the market which you can choose from, and you can buy the cheaper one. The cheaper ones are going to be more energy efficient and more kind to the environment than the expensive ones. But [energy] is not a product that you can choose. I mean, everybody has to pay twice as much. That's not really a market. How is that a market?—Marge (lawyer, environmentalist)

This is a clear example of the difficulty in understanding how energy efficiency can be substituted for fuel. And Marge is an environmental lawyer, an intelligent person who reads environmental newsletters and policy papers and who advocates "market solutions" to environmental problems.

Part of this problem is that people do not see energy efficiency as an option nor understand its ramifications. But there is some further block, because even Nick, the informant who most clearly understood energy efficiency, had the same reaction. Notice that we have to explicitly mention efficiency as a reaction to the fuel tax, and when he finally makes the connection, he reacts with surprise, despite "already knowing" the relevant facts.

> I'd be pissed off. . . . Everybody drives to work and back [long discussion about the unfairness that poor people will pay a larger proportion of their income] . . . *Some people will claim that if it costs more to buy fuel, then people will use less fuel.* Drive halfway to work? *Right* [*laugh*] . . . *Or have the extra incentive to buy a truly fuel efficient car.* Well, that's true [spoken slowly, with emphasis]. I mean, that certainly did work. *When?* In the seventies. *So the total outlay is the same, it just happens that* [*less fuel*] *is going through a more efficient engine.* I just think the support will not be there if you hit people like that. . . . The whole cost of research could be shuffled out of unnecessary parts of the federal budget and nobody would know the difference.—Nick (small manufacturing plant owner)

Nick's claim about the seventies is correct. When gasoline prices increased, Americans bought much more efficient cars rather than reducing miles driven (Greene 1987). Nick is an intelligent man who has an exceptionally clear understanding of energy efficiency. Nick already knew that consumers responded to earlier gasoline price increases by buying more efficient cars. Why was he surprised when we mentioned that the same thing might happen again?

From Nick's and Marge's arguments, we hypothesize that, even when people understand energy efficiency, the other problem is that "taxes" are associated with an entirely different realm of discourse. That discourse concerns effects of tax policy across income strata and service-using groups, punitive versus fair taxes, limitations of personal monthly budgets, and concern about government's ability to spend wisely.

In our survey, the pattern of responses to fuel taxes was complex, providing some insights into the internal structure of what prior public discussions have regarded as a monolithic opposition to this policy. We outline our argument by stepping through the relevant survey statements.

First, substantial majorities accept environmental taxes in general.

23 People should pay the environmental costs of the things they buy. Products should be taxed depending on their effect on the environment.

Earth First!	Sierra Club	Public	Dry cleaners	Sawmill workers
97	85	70	77	48

Most accept the economic logic of fuel taxes in particular, with agreement dropping well below majority only among the sawmill workers.

42 Energy conservation will work better if we price energy correctly through higher fuel taxes to make efficient energy use in people's own interest.

Earth First!	Sierra Club	Public	Dry cleaners	Sawmill workers
100	76	73	50	30

Apart from the environmentalists, survey respondents are about equally split on whether their own fuel use is elastic.

20 I would still buy the same amount of gasoline no matter how high the price went, it's a necessity.

Earth First!	Sierra Club	Public	Dry cleaners	Sawmill workers
16	37	53	60	44

Agreement is higher when the statement claims only that there are some necessities "such as heat" with inelastic demand.

138 There are certain basic necessities, such as heat, that people cannot give up no matter what the environmental costs are.

Earth First!	Sierra Club	Public	Dry cleaners	Sawmill workers
32	63	76	70	67

The "necessities" view leads all groups but environmentalists to view fuel taxes as potentially unfair.

92 A fuel tax would be an unfair way to reduce fuel consumption because some people are forced to use more fuel than others by their business or personal needs.

Earth First!	Sierra Club	Public	Dry cleaners	Sawmill workers
17	33	70	67	81

Notice that the subtle wording difference from statement 42 to statement 92 shifts response from 70 percent support to 70 percent opposition. Notice also that in the above series of survey statements, environmentalists have a view closest to economists and efficiency analysts (and the experience of the seventies)—that reducing fuel use is possible and is a likely reaction of consumers to higher prices.

National polling also supports our conclusion that energy taxes are opposed due to specific cultural models rather than resistance to taxes in general. In February 1993, a Yankelovich survey found majority support for new taxes to reduce the deficit, with support for specific taxes varying widely: 75 percent supported taxes on tobacco and alcohol, 32 percent supported taxes on health benefits, and only 23 percent supported taxes on energy (Church 1993). (An energy analyst might argue that energy taxes should be supported as much as tobacco and alcohol taxes, since all discourage consumption of a product that causes social harm even to nonusers.)

In contrast to the Yankelovich survey, a Greenberg-Lake survey at the same time (January 1993) presented voters with a more completely argued proposal for energy taxes. Their question proposed to "discourage pollution and energy inefficiency by instituting taxes on pollution and energy use and using that money to reduce the deficit," and 71 percent supported it (Greenberg-Lake and Tarrance 1993). This big jump—from 23 to 71 percent support for energy taxes—is consistent with other polls showing substantial majorities supporting taxes to reduce pollution and with our own research showing that voters do not yet recognize the environmental benefits of energy taxes if they are not stated explicitly.

In summary, our energy tax proposals were opposed because of a strong belief that fuel consumption cannot be changed (it is inelastic) and that current levels are a necessity, both of which make the tax seem unfair. Further, reactions to the gasoline tax focused on miles traveled rather than gallons used ("Drive halfway to work?"), a view related to the inability to conceptualize energy efficiency noted in the preceding section. We also hypothesized what might be called a "realm of discourse" effect: as a tax proposal, the energy tax was evaluated on the basis of criteria such as impacts on different income levels, whether the tax could affect demand, and the likelihood that government would use the money wisely. Tax proposals are normally not evaluated on criteria such as environmental issues, economic incentives to conserve, or energy efficiency.

Returning to the Clinton Btu tax battle, the opposition seems to have figured out some of the reasons why the public reacts negatively to energy taxes. One advertisement against the Btu tax stated: "Alarm clocks will cost more, hot water will cost more and the breakfast we eat will cost more because it takes energy to make them or get them to you. . . . Quite literally, take a shower and pay a tax. Think about it."[7] This assertion stresses the beliefs about inelasticity of energy and unfairness of taxing necessities, which we found made people oppose energy taxes. Equally appealing (and less misleading) arguments could have been made in favor, but the Clinton administration did not try to do so, and there was no well-funded support for the Btu tax to match its opposition.

Models of Institutions

Americans' preferences for policies to resolve environmental problems also depend upon their views of the mediating institutions that may help cause, detect, or prevent global warming. The institutions most related to these three aspects of global environmental change are science, industry, and government. Previous studies have shown that on environmental matters the public trusts scientists most, followed by government, and trusts industry least (e.g., Milbrath 1984: 128). This section describes some of the reasons for this ranking.

Science

Because global-scale environmental change cannot yet be seen directly, citizens must rely on scientists' measurements, theories, and predictions to know about it. The degree of public confidence in the scientific community is therefore crucial, as is the mechanism for transferring that information. Prior studies have shown that scientists may be mistrusted in environmental controversies, especially involving local risks. The science establishment itself often attributes this mistrust to the public's lack of scientific understanding (Collins 1987; Wynne 1991). For example, case studies show scientists may be mistrusted when they refuse to acknowledge disagreement and mistakes within their own ranks, display ignorance of local mediating conditions, or seem to be serving as apologists for distrusted institutions (Wynne 1992). Only the first item on this list would seem to apply to global environmental change, in which no local community is involved and the national media, rather than local hearings, are the primary communications channel.

We find many expressions of confidence in independent scientists regarding environmental matters. This was expressed in the semistructured interviews and in the survey. For example, people denied that they needed to see a problem personally to be concerned about it.

106 Just because something is in the newspaper or on TV doesn't mean it is something to be concerned about. To be concerned, we have to see a problem with our own eyes.

Earth First!	Sierra Club	Public	Dry cleaners	Sawmill workers
40	19	30	20	33

Note that the radical Earth First! members are most skeptical of problems they do not observe personally. The answers to the above question suggest that even unseen problems can be politically important, given a trusted intermediary such as the scientific community.

Furthermore, our interviewees also felt that they could reach their own conclusions without total consensus among scientists. Majorities felt this way in all but one of the groups.

46 The so-called "experts" on global climate change disagree so much with each other that it is hard to know what to believe. We really don't know whether there will be global climate change or not.

Earth First!	Sierra Club	Public	Dry cleaners	Sawmill workers
10	30	33	37	52

In the semistructured interviews, some of the congressional staff questioned the credentials of the most publicly visible scientists. This questioning of credibility was expressed by a minority of the public in our survey (although a substantial minority).

137 The scientists who get into the newspapers and on TV are not in the mainstream with other scientists. They are on one fringe or the other.

Earth First!	Sierra Club	Public	Dry cleaners	Sawmill workers
23	11	40	20	48

We speculate that one of the reasons we find higher trust in scientists than have studies of local environmental risk communications may be that there are really two institutions legitimizing the information; the media serve somewhat like peer reviewers. That is, citizens know that not only has a scientist made the claim but also that the media have decided to print or broadcast it.

The public's trust seems to be based on three assumptions: scientists do not release findings to the public without evidence and deliberation, the media do not publicize scientific findings that are not accepted and significant, and independent scientists (not working for industry) have no reason to lie. We hope that our finding of public confidence can be a reminder to scientists that they have a valuable credit with the public that they must continue to husband if it is to be available when needed in the future.

In an indirect validation of our analysis of the trust in scientists, we note that Limbaugh's (1993) argument that ozone depletion is exaggerated includes an attack on scientists, claiming that they have exaggerated dangers of ozone depletion in order to get more research grants. This might otherwise seem to be an ad hominen attack, subordinate to his main argument, but by our analysis it is essential to his argument because it breaks the otherwise strong credibility of independent scientists on environmental matters.

Industry

The second environmentally significant institution, industry, was seen as the major source of pollution and as financially motivated to resist environmental protection. This combination led to considerable mistrust. For example, the "wait for evidence" policy discussed earlier was widely attributed to vested interests.

19 It's the people who have a vested interest in exploiting the environment who say we have to wait for more studies of global climate change before we do something.

Earth First!	Sierra Club	Public	Dry cleaners	Sawmill workers
100	89	82	67	63

One indicator of the distrust of industry was the astonishing 80 to 90 percent of the public believing that automobile efficiency improvements have been blocked by a conspiracy between the automobile manufacturers and big oil companies. People feel the profit motive blinds industry to human concerns.

I don't believe this is a disposable world, even though a lot of people feel that way. *Who do you think might feel that way?* . . . Say, people that are very narrowminded, and don't really see the beauty in the world that nature has to offer. They might think more of a profit. I think it might be somebody that's . . . [for example] in the chemical industry, who's . . . very money oriented.—Doug (pharmaceutical scientist)

This leads to the resulting belief that industry has not done all it can for the environment (public relations campaigns notwithstanding).

70 Corporations and utilities have just about reached the end of what they are able to do for the environment.

Earth First!	Sierra Club	Public	Dry cleaners	Sawmill workers
16	7	17	17	8

This belief is consistent across our five groups. The cultural model explaining poor corporate environmentalism was not that corporate officers were bad, but that industry is entirely directed by the profit incentive. As an informant quoted below put it, in an explanatory rather than accusing tone, "They are motivated by greed, which is the nature of the beast."

Government

The third institution, government, is viewed as a "necessary evil" in environmental matters. Given the suspicion about industry, most informants saw government as the only institution capable of controlling industry.

Do you think government should get involved in regulating the corporations? It's obvious that the industry won't regulate itself when it comes to [pollution regulations] because of the cost and so on. Someone's going to have to do it, and I don't know who else would.—Bert (resort proprietor, hunting guide)

Corporations won't do it on their own. They are motivated by greed, which is the nature of the beast. There would have to be government regulations. . . . Maybe international bodies regulating the environment. It would probably take some disaster before it's ever come into effect in the foreseeable future.—Charles (coal mine construction)

A few informants took a more optimistic position, assuming that industry is motivated to take environmental actions without government involvement.

I think [companies] would naturally want to [adopt clean, efficient new technologies for factories]. *Why would they?* Because if it's safe and effective, and seemingly more efficient, they would want to do it to save their own costs in the long run and to set the example, to make their mark. Cause pretty soon, and even to some extent already, people are boycotting companies that aren't environmentally sound in their production, and that's just going to increase until companies react.—Kate (college student)

The survey showed that the vast majority of our respondents expect neither consumer-driven nor industry-internal motivations to be sufficient.

89 Unfortunately a lot of companies wouldn't do anything to protect the environment unless they were forced to by law.

Earth First!	Sierra Club	Public	Dry cleaners	Sawmill workers
97	100	97	93	78

But concerns about government lead to very mixed feelings, to a view of government as a necessary evil.

It [government involvement in combatting the greenhouse effect] is a necessary evil. *What are the costs?* . . . Well, it's going to hurt everybody financially. You may not have as many freedoms as you have now. It's going to create more bureaucracy. . . . Power tends to corrupt. Those are the drawbacks to whatever is established.—Charles (coal mine construction)

A minority has such a loss of faith in the government that they want to turn everything over to the private sector. One such person, Susan, said that she would not support a tax on energy sources producing greenhouse gases. We then used the follow-up question noted earlier, asking whether her opinion would change if the tax were used to fund research on new energy sources.

I don't trust the government. They're not going to spend 100 percent of my money on research; first there'll be an 80 percent administrative fee. I think it should be private industry, not the government. It would be more closely monitored if it were private industry, not the government.—Susan (hospital administrator)

In the collection of tax revenues, the government can enter the same self-interested domain usually inhabited by industry. This is yet another reason in addition to those mentioned earlier that energy taxes are problematic.

The institutions of government and industry were not seen as independent. For example, large majorities of respondents attributed government failures in environmental regulation to industry.

55 The reason politicians break their promises to the people to clean up our environment is the power of industry lobbies.

Earth First!	Sierra Club	Public	Dry cleaners	Sawmill workers
97	89	90	90	85

Similarly, many Americans also infer that there is some manipulation of information by industry and government.

We're at the mercy of what they want to tell us on the TV news. It's canned, not unlike what people say news is like in the Soviet Union; it's censored. *Censored by the companies or the government?* By the companies. Also to some extent by the government, I'm sure, what they want us to hear.—Ellen (freelance writer)

We originally considered this a fringe position, almost paranoid. However, when placed on the survey, it received majority agreement from four of our five groups.

56 The news the public gets on the environment is censored by the big companies and the government.

Earth First!	Sierra Club	Public	Dry cleaners	Sawmill workers
94	41	77	77	59

To summarize the views about government, it was seen on the one hand as inefficient, unresponsive to citizens, frequently corrupted by industry lobbying, and driven by internal demands, especially maximizing tax revenues to preserve bureaucratic jobs. As Bert put it, "this government's not designed to save the planet." On the other hand, government is seen as the only institution powerful enough to control industry. Although government regulations were seen as leading to inefficiency and red tape, regulations in general were thought to do more good than harm, and this belief was even more emphatic for environmental regulations in particular (see our survey questions 81, 148, and 149).

We have only scratched the surface of cultural models of institutions related to global environmental change. These initial explorations convince us that there would be much interest in a fuller investigation of the cultural models with which laypeople understand science, industry, and government. We have not touched on other relevant institutions that play important intermediary roles, including environmental groups, the media, and interpersonal communications channels.

Yet even our preliminary investigations help us begin to understand some puzzles. For example, our data clarify why Americans believe scientists' claims about invisible global phenomena; respect industry's competence while distrusting its motives and environmental claims, at times almost to a paranoid degree; and dislike governmental regulation yet favor environmental regulations on industry.

Conclusions

Many of our findings regarding policy preferences can be explained by cultural models, beliefs, and values. When asked, prior to our briefing, for a list of policies to prevent global warming, laypeople produced a list with few effective policies, but predictable from the cultural models described in chapter 4. Specialists were less likely to suggest ineffective policies based on cultural models. However, in part because they have different sources of information, specialists from different sectors differ greatly in the aspects of the problem and possible solutions they could produce in response to our open-ended questions.

Lifestyle changes were seen as one effective way to deal with environmental problems in general and global warming in particular. Majorities expected a need to reduce our "level of consumption" but did not expect those reductions would reduce their "standard of living." Most expected a need for both lifestyle changes and new technology. Surprisingly, substantial majorities of all groups approved of "forcing" people to change lifestyle for environmental reasons.

Laypeople had a positive attitude toward energy efficiency and supported the one specific efficiency policy in the interviews—doubling of automobile MPG standards. However, energy efficiency was very poorly understood, and was rarely mentioned as a solution to global warming. Majorities in our survey incorrectly believed "reducing pollution" is more effective than energy efficiency for dealing with global warming.

The conceptual problems are compounded in the reactions to energy taxes. Although majorities supported environmental taxes in general, energy taxes were not seen in this category. We analytically separated out the components of opposition to energy taxes (for example, energy as a necessity and unchangeable, energy tax as regressive and not affecting behavior), finding specific beliefs or models as primary explanations for opposition rather than selfishness or a general opposition to taxes.

How might these findings be applied? As examples, we consider advocates for two types of policies, energy efficiency and a fossil fuel tax.

Efficiency has the initial advantage that people have general positive associations with it. However, when efficiency policies are contested in the political arena, efficiency advocates would need to target the existing barriers to deeper support, beginning with linking energy efficiency to environmental benefits. They would need to explain why, in issues like global warming, energy efficiency is more effective than pollution controls, not less. For policies that deliver efficiency to the consumer, they would need to watch for the tendency to associate lower energy consumption with diminished services and try to overcome the difficulty people have in conceiving the core concept of efficiency—that one actually can achieve the same output from less energy input. For policies motivating manufacturers to make efficient equipment, the challenge will be to explain why manufacturers are not already maximizing efficiency, without resorting to the appealing conspiracy theories.

The case of Nick, the furniture shop owner-operator, shows that a well-measured familiar example can dramatically improve understanding of efficiency and increase support for subsequent energy efficiency policies. This case suggests a strategy of providing good, clear examples of existing efficiency improvements. What examples could efficiency advocates use? Utilities promoting efficiency are now distributing compact fluorescent bulbs, which produce light using about one-fourth the electricity of standard bulbs—a huge savings. But the problem with this example is that the savings are not visible—residences do not meter individual light bulbs, and the actual savings are lost in random fluctuations of the monthly bill for the whole household. Another candidate example would be increased automobile efficiency. However, as Nick's case shows, even when individuals are familiar with this example, they may not generalize it to other cases of energy efficiency.

The public communications task would be more difficult, but by no means impossible, for proponents of a fossil fuel tax. A simple start would be to frame the fossil fuel tax as a pollution tax, which one pair of our survey questions suggests can flip opinion from 70 percent opposition to 70 percent support. Advocates of energy taxes also need to break the view of energy as an unchangeable necessity; one method might be to point out what happened with higher gasoline prices—U.S. automobile efficiency doubled, holding down fuel use even though people drove more. This suggests a legislative package that would be sound policy and would simultaneously help make conceptual connections to increase public support: combine a gradually increasing energy tax with energy efficiency regulations to reduce use. Additionally, a fraction of the revenue could be spent to promote renewable energy and to compensate for the income regressiveness of a consumption tax.

We do not propose suggestions that take advantage of misconceptions or improper use of cultural models. For example, we do not advise that energy efficiency advocates attribute currently slow MPG progress to a conspiracy of automobile manufacturers with oil companies, any more than we approve that opponents of the Btu tax exploited the view of energy use as unchangeable. We return to the question of potential for misuse of our findings in chapter 9.

7

Case Studies of Influential Specialists

This chapter describes individual informants, drawing from the semi-structured interviews, with three goals in mind. First, the cases show how beliefs and values are connected by individuals and applied to real decisions. Second, these cases, combined with the analysis in the next chapter, lead to a more integrated view of the components of environmentalism. Our third goal is to show how individuals choose and selectively remember information, enhancing differences of opinion among them.

Each of the people in these case studies is actively involved with U.S. environmental policy. They include two congressional staff members and two advocates from nongovernment organizations (an environmental organizer and a representative of mining union interests).[1] One of the congressional staff and one of the advocates works for environmental protection, the other in each category often opposes it.

The assimilation and use of information by these participants is influenced by their own mental models, values, and political ideology. These specialists are nonscientists who employ scientific results in their work. We find several instances in which these individuals actively search for information that fits the constraints imposed by their ideology and their employer (whether an industry, a congressperson, or an environmental group). As a result, these individuals differ on basic facts. To illustrate information flow, we trace one example of environmental information moving from a scientific journal, through news reporting, to a memo for members of Congress—undergoing substantial modifications along the way.

Each of the four cases begins with background on the person and the context of the interview, omitting details specific enough for positive identification (even though none of these four requested anonymity). Our description of each person's views emphasizes the areas they expanded most upon, in keeping with the semistructured interviews themselves.

Why Pick Specialists?

This chapter focuses on the views of specialists for several reasons. As a practical matter, the people in this chapter are our closest approximation to those who will actually make our society's decisions on environmental policies—environmental policy decisions are made not by scientists but by people who interpret and use their work. Since we have a representative government, decisions are not made by citizens operating purely on media coverage. Although legislators cast votes, legislators rely heavily on advice from staff like those described here, who assess constituent concerns and collect predigested information from interest groups, like the advocates in the two case studies. All parties also try to influence public opinion. Thus, our practical reason for picking these specialists is that they will have a disproportionately large effect on what happens in environmental policy.

A second reason for picking specialists is a sampling strategy. Although they may not follow all the details as a scientist would, they should be well informed on the ramifications and policy implications of scientific knowledge. In areas related to legislation or policy, they should be less likely to apply cultural models inappropriately. As noted in the previous chapter, this was generally the case. Of our four cases in this chapter, three volunteered so much of the information planned for our briefing that we dropped the briefing or mentioned only the facts they had not offered themselves. For the same reasons, these specialists are, of course, not typical citizens. For comparative purposes, appendix D presents three case studies of laypeople, not specialists.

Congressional Staff

Five congressional staff were interviewed in Washington, D.C. All worked on environmental or energy issues as their primary specialty. They were picked in advance to represent both those supporting and opposing immediate action to limit global warming. We present one case study from each side: Alvin supporting action and Gerard opposing it. They are fairly balanced in their background on these issues. Both are knowledgeable about environmental law and policy but have other responsibilities as well. Also, both are staff of members of congress, not committee staff, so they have spent a few years, not decades, on environmental policy.

Alvin (legislative aide)

Alvin is a legislative assistant for a Democratic representative from an area with high environmental concerns. In our first telephone contact, prior to the interview, we said we wanted to ask Alvin about his views on global warming. He responded: "It's important, and we need to do something immediately." Alvin had already drafted a bill, introduced by his congressperson, that would have taxed the carbon content of fuels in order to provide an economic incentive for reducing CO_2 emissions.

To conduct the interview, Boster, Kempton, and a research assistant (Leslie Clark) had an appointment with Alvin at his Capital Hill office. We waited as he went past the scheduled time in a meeting with coal industry and mine worker representatives (the "coal boys," as he referred to them later). After a half-hour wait, another legislative aide passing through the reception area reported to us: "He's getting beat up real bad in there." (Alvin's proposed bill would tax all fossil fuels, but because coal has the highest carbon content per unit of energy, a carbon tax would place coal at a relative disadvantage against other fuels.) After fifty minutes, seven men filed out of his office, talking and laughing among themselves.

Perhaps because of his confrontational prior meeting, our interview was strained, and Alvin's answers were terse and somewhat hostile. He refused our request for permission to use the tape recorder—the only one in the forty-six semistructured interviews to refuse—apparently for fear of being quoted out of context. Therefore, our quotes here are based on notes taken during the interview.

Consistent with our findings in the chapter on values, Alvin argues for environmental protection first based on utilitarian values, saying that it is necessary for health, and mentioning early in the interview that pollution costs companies in health benefits for sick workers. More generally, he says, it does not make sense to poison your own nest. Like others mentioned in our discussion of values, Alvin thinks contact with nature increases environmental awareness. He feels the relationship between humans and nature is "integral and bound together," but that today the relationship has "disintegrated." "The average person thinks the pork chop is made in the supermarket." He comments that if people do not understand something, they tend not to care about it, elaborating:

> [People] ought to experience the natural world. Even if it's growing a patch of grass in a New York tenement, so that people get a sense of connection with the environment. It's fun to read about it, but you must go a step beyond. You could go to the zoo. But it would be better to grow a patch of grass . . . you have the experience of planting it, watching it grow.—Alvin (legislative aide)

Alvin believes he has strong environmental values, saying "you always leave the campsite as you found it," an expression he attributes to the Boy Scouts. When asked who that obligation is to, he says the obligation is as much to the environment itself as to other people.

Alvin gives a fairly complete description of global warming, mentioning CO_2, methane, and CFC as source gasses. Regarding effects, he states that it would "radically alter climate." He mentions chaos theory to make the point that the effects may be unpredictable, saying that North Dakota could be growing mangoes instead of winter wheat.[2] He also mentions icecaps melting and raising sea level, increased intensity and destructiveness of storms, decreased food production, and brownouts from utilities not being able to meet summer air conditioning loads.

When we later presented our list of potential effects, Alvin became very angry because he thought our list was much too weak.[3] He then launched into a scolding tirade, rebutting several items from our list of effects.

> The way you have phrased these sounds trivial. When you say "slightly higher food costs," that doesn't begin to define the problem. You're talking about

[climate change having] a major impact on agriculture, like the desertification and famine in Africa. It's not just, "some animals and plants move," like fire ants moving north; you are losing major species. . . . Plants and animals can't move that fast. We'd be losing important hardwood trees. It would be a significant coastal inundation, not just a random beach cottage going under.

As he continues, Alvin gives a clue about his motivation for choosing the most disastrous predictions.

If we are going to impose a tax that would have a 30 billion dollar cost in five years' time, we'd better have a more serious justification for it. . . . [Bangladesh] would be not just damaged but wiped out. What is the cost of flooding Manhattan? There is a big degree of difference here that you just don't address. This [the briefing's list of effects] is bullshit. . . . I'm talking to the coal boys asking them to take a big hit and give up their jobs. This list is not a reason to significantly alter industrial policy.

Alvin gets information on global warming from newspapers, magazines, scientific conferences, and "the environmental community." He does not read scientific journals, commenting:

My job is in the political arena. I believe the science, I don't need to understand all the details of the chemistry. . . . You can look at people's credentials to trust them, its pretty clear on scientific issues.

We asked how he evaluated conflicting scientific claims.

If I read a study and then find the study was funded by the coal industry, that's going to affect my evaluation. . . . Also you can look at people's past stands and evaluate them. DuPont maintained, starting in 1972, that CFCs didn't damage the ozone layer. . . . Intuitively, global warming makes sense. If you've got an atmosphere and you're burning hot stuff, it makes sense that it's going to get hot.[4]

Alvin had strong reactions to our list of policy options. He dismissed the option of adaptation without prevention, saying: "That is obviously not a serious proposal." His reaction to fuel efficiency regulations was as critical and more sweeping.

Regulation of the auto is penny ante. The individual policy [for individual products] is penny ante. You need to fundamentally alter the way energy is priced.

Because Alvin wrote a carbon tax bill, we were not surprised that he liked the interview's proposed policy option for fuel taxes. Reacting to our hypothetical 100 percent tax, Alvin admitted that, as set then, his carbon tax was much too small to affect fuel consumption significantly, but he felt that his bill would put a mechanism in place that would easily be raised in the future: "A 100 percent level is not politically doable." It is illustrative to compare Alvin with Mark, the unusually knowledgeable congressional staff member described in chapter 6. Mark was also working on a global warming bill, one that contains many specific regulatory and tax incentives to reduce greenhouse gas emissions. Alvin does not read the technical literature or know energy efficiency options to nearly the same extent as Mark. Thus, it is convenient for Alvin to propose a carbon tax rather than regulation because his bill could have equally broad effects, while not requiring of him substantive background knowledge.

Consistent with the typical environmentalist position, Alvin did not like nuclear power because of the waste disposal problem and the possibility of catastrophic accidents. On the question of new technology versus changing lifestyles, he says "I don't think you need to fundamentally alter the way you live."

After the interview, Alvin asked what motivated us to do the study. We answered that it was of academic interest to us, that research and publication were part of our university jobs, and that we hoped to improve public understanding of global warming. He replied that public understanding was already sufficient, offering several examples.

> I would say the whole idea of global warming is generally accepted, as is the ozone hole. I've even heard it used by a sports announcer. When they're using it, you know it is widely known. The public deals with global warming in the same way they do problems in Latin American countries. They don't know Guatemala from other countries, they just know we're supporting something down there that they don't like, and they want to stop it. In this case, it trickles down that opinion leaders think global warming is a problem. People don't need to know all the details. If we just pass the carbon tax, that will take care of it and we'll move on to the next problem.

In summary, Alvin is a nonscientist who has mastered enough facts of the global warming controversy to do his job without directly reading the scientific literature or understanding all the physical principles involved. Alvin has embraced one policy, the carbon tax, to the exclusion of others, perhaps because it is consistent with the current preference for economic incentives rather than "command-and-control" regulations. He expects this policy to have undesirable economic costs, so he selects and emphasizes information to construct global warming as being very destructive and dangerous. Whether or not he is right that our list of effects is too weak, he became highly agitated—not because of a detached passion for scientific accuracy but because less dramatic effects undermined arguments for his proposed legislation. Having committed to a single policy, he has focused his orientation to new information on building arguments to support it, a task whose importance is reinforced by getting "beaten up" periodically.

Gerard (legislative counsel)

Gerard is a legislative council, handling health and environment legislation for a conservative Republican congressman. He has a law degree and describes himself politically as conservative on economics and foreign policy, middle-of-the-road on social issues. He is articulate, sympathetic, and has a good sense of humor. Boster, Kempton, and Clark interviewed him in his congressional office.

Gerard justifies environmental protection for health reasons and for the economic value of recreation. When asked about environmental values, he mentions his enjoyment of the beach and his anger at seeing "things floating in the ocean that aren't supposed to be there." On the other hand, he views circumstances such as the huge number of cars in the United States as the price we must pay "for the economic freedom and progress that we have in this country." Nevertheless he favors mass transit and incentives for getting people out of their cars.

Despite his initial anthropocentric justification for environmental protection, when we later asked Gerard for his reaction to our list of possible effects of global warming, he singled out species extinction as one of his two primary concerns. He picked it because we have "an obligation" to preserve species from extinction. When asked who the obligation is to, he said "To nature, to the whole ecosystem, to the reason we're all here . . . there's a good reason to have an Endangered Species Act." Gerard's views on the unpredictability of nature's interdependencies and the obligations to preserve species—strong proenvironmental arguments in each case—are cited in the chapters on models of nature and values (chapters 3 and 5).

Recall that in our previous case study, Alvin could not think of any examples to answer our question about environmental regulations going too far. By contrast, Gerard had an extensive answer.

> The perception that a lot of us have around here is that politics drives the environmental issues in a really fundamental way. . . . And the environmental organizations have much more powerful and motivated grass roots organizations. *Than industry?* Than industry would, or even just the silent majority, wherever they may be. And the media seems to have a certain bent on this issue.

Gerard sees a political process almost under siege by exaggerated but potent environmental concerns.

> You will find city council meetings in little townships anywhere along the coast . . . where they'll be inundated with 200 people demanding the city council take a position on offshore oil. Now their position on offshore oil is technically irrelevant. . . . But they'll do it, and they'll send that resolution to their representatives in Congress. And that makes a difference. And [members of Congress] are getting inundated with . . . this grassroots concern. There's no countervailing group of people who depend for their livelihood on this, or who might think that there's environmental advantages to offshore oil, also going to city councils.

Gerard says that if he explains his positions to "average" citizens, the typical response is that it makes some sense, but they hadn't thought of it. They wouldn't have thought of it, Gerard goes on, because "it wasn't on any of the news reports. It doesn't seem to crack newspapers." He describes what he sees as a result of one-sided reporting and one-sided community activism.

> And the next you know you're driving a 10 million ton SO_2 reduction bill through the Congress. Whether or not 10 million tons is different than 8 million or 6 million, no one can tell you that, technically. But 10 million was a nice round number, we chose it, and even if it means disadvantaging certain entire utility systems, so be it.

On the issue of global warming, he suggests that this issue is being used for "other purposes": to raise money for environmental groups, to win an election, or to influence a partisan ballot initiative.

One result of this exaggerated concern for the environment is increased cost of products. Buyers, however, do not see those costs itemized on the sticker price. Gerard mentions a hypothetical figure of a 20 percent price increase. (We note that the actual average price effect of environmental regulations is less than one-tenth that, or 1.5 percent.)[5] He asks rhetorically "What additional benefits do you receive from that 20 percent add-on cost of this product?"

> You have a theoretical advantage based on some models that we've used to assess the risk of toxic air pollutants to the maximum exposed individual who's setting buck naked on top of the factory stack for 70 years, 24 hours a day. [interviewers laugh] That's the standard they use, the maximum exposed individual . . . it usually means factory gate, arguably it could mean on top of the smokestack.

Gerard may have some legitimate criticisms of how science policy determines what is safe. Nevertheless, his use of an absurd and humorous image is clearly part of a strategy of persuasion. In response to our laughter, he said:

> You laughed, but when you explain that to the average person . . . that you're going to shut down this plant for some egghead sitting in front of a computer who's churning out numbers that have no relation to reality that we know of.

Gerard says that if more voters understood the scientific basis of environmental regulations (as he sees them), voters would still have "a strong environmental concern," but it would be "a more targeted and intelligent one."

Gerard gives a reasonable summary of what global warming is about.[6] The crucial area in which he differs from proponents of action like Alvin is in his judgment of the certainty of the scientific evidence.

> The range of expert opinion is so broad on this, it leads to political manipulation on the Hill. . . . When you're trying to make policy, you're trying to address something in a [scientifically] certain framework, you're used to dealing with what's the mainstream [expert] opinion and what's not. . . . On this it seems like there's no mainstream, there's no fringes, there are just people all over the lot.

Consistent with his earlier statements about environmental issues in general, Gerard suspects political manipulation of the global warming issue. He sees the claim of a scientific consensus that warming will occur as being made by politicians who favor action and the left-leaning press. He does cite one "pleasant exception" to this bias.

> One news show I watched where a fellow from the University of Virginia [probably Patrick J. Michaels or S. Fred Singer] was debating Michael Oppenheimer [of the Environmental Defense Fund]. . . . It came out pretty clearly to

anyone who was watching that there is a range of disagreement on this, and two people with equally impressive credentials can disagree.

Given this range of opinion, how does Gerard decide what to do? We asked him how he chose from alternatives of conflicting expert opinion. His methods are similar to Alvin's.

Let's say Senator Wirth and [Representative] Claudine Schneider are making a lot of noise about global warming. And if there's a wide range of [scientific] opinion, you check out what is your opinion of Schneider and Wirth. And if they're always on one side of these issues, you sort of discount what they're saying to some degree. Similarly, if people in the press or media are writing about it . . . who do you trust on other things.

Gerard uses similar criteria in evaluating information from scientists. He says that if conclusions are based on "a model put together by a scientist who may have . . . [a] political side that we don't trust, we start to pull back a little bit from the conclusions." Some scientists may wince at this method for choosing among conflicting expert opinions, but given that neither the media nor the scientific community has provided a clear signal, and given limited time and evaluative capabilities, Gerard's method is arguably a reasonable general-purpose decision strategy. For staff such as Gerard and Alvin, who must survive in the political arena, it may be a general strategy used to evaluate information on other issues in addition to scientific ones. This method of evaluating science for legislative decisions differs considerably from the procedures followed in official channels such as the Office of Technology Assessment or the Congressional Research Service.

Gerard also evaluates scientific claims on the basis of the consistency with other positions of their proponents. For example, he uses nuclear power, which emits no greenhouse gases, as a litmus test.

A lot of the people who are ringing the alarm bells on global warming are also the most likely to be antinuke people. And, if they really feel that strongly about global warming, one would argue they should be more willing to tolerate an expansion of the use of nuclear energy. And that doesn't seem to be in the offing, although I'm told that there's beginning to [be] a little schism in the environmental community on this issue. And that's good, that's healthy. Because that tells me . . . that people do take the [global warming] issue seriously and believe it's going to happen. And they're not just trying to manipulate the political system to generate more votes for the environmental issue or more money for environmental organizations.

Given Gerard's assessment of political biases on this issue, it is not surprising that he is receptive when he finds a "more recent . . . NASA study that showed the temperatures pretty much going back and forth over time." This evidence came to Gerard's attention via an article in *The Washington Times,* which in turn cited *Science.* Gerard seized this information both because it was a counterbalance to what he sees as the prevailing bias of the media, and because he is skeptical of modeled (versus measured) data.[7] Better than models, he felt, would be something like "the NASA study, where you actually have hard data that goes back fifty years."

Because this "NASA study" fitted so centrally with Gerard's strategy, he wrote a memo for his representative to circulate as a "Dear Colleague" letter (a memorandum from one representative to the entire House on a topic relevant to pending legislation). At the top of his letter was the graph from the *Washington Times.*

We trace the history of this graph briefly because it illuminates how specialists select and modify information and are in turn recipients of preselected data. The graph, which originated from a scientific journal (Spencer and Christy 1990, 1560, fig. 4), showed ten years of temperature data for the United States, comparing ground station measurements with satellite temperature measurements. The graph demonstrated that the two lines were equivalent, that is, that satellite data could be used worldwide—where it could not be corroborated with surface measurements—in the future to establish whether global warming was occurring. In the global data, the authors note that no trend was ascertainable, but that the ten-year span of data would be insufficient to see a trend amid normal fluctuations. The *Washington Times* news story on the Spencer and Christy article, 30 March 1990, was headlined "NASA satellites find no sign of 'greenhouse' warming." The news article included the U.S. temperature graph, reworked by the newspaper's art department to include a large caption "No proof of global warming" and a label "Satellite global temperature readings." Despite the misleading headline and label, the news article does quote one of the original authors who stated "This does not prove that there is not a global warming." This clarification is lost in Gerard's "Dear Colleague" to the House, which said "The NASA study offers strong evidence that there is *no* 'greenhouse effect'" and concluded "environmental extremists are making an issue out of several unusually warm years in the 1980s to solve a problem that, as the NASA study shows, does not exist."

Our point in tracing the history of the graph is not to find fault with this particular newspaper or staff member—similar examples could surely be found from the proenvironmental side. Rather, our point is to illustrate how information is selected and transformed as it moves from the scientific, through the journalistic, to the political realm. The treatment of data is also that of a lawyer supporting his client's case, rather than that of a scientist's careful weighing of the conflicting evidence. Even if Gerard's letter did not convince anyone on the opposite side, it may have been very important for providing scientific ammunition for those Representatives already inclined to oppose Congressional action on global warming.

Gerard sees the current scare over global warming as just one more example of "theoretical risks," which he has heard about since he had to read Erlich's book, *The Population Bomb* (1968), in elementary school. Gerard recalls being afraid when he read about "years like 1990" when Erlich predicted "400 billion people . . . all killing each other for food and eating each other." The elementary school teacher who assigned the Erlich book also suggested that the students clean up the street in front of the school for Earth Day. This type of activity is a common school exercise. However, it suggests that litter is a more serious concern on Earth Day than, say, habitat destruction, and suggests that environmental problems are solved by individual voluntary actions rather than collective institutions. The class did clean the street, at considerable effort. Gerard recalled that the day after the cleanup, the street was as dirty as before. He used this example to make the point that environmental cleanup, in general, often was not worth the effort. Thus did a scientist's overblown population predictions and a teacher's trivialization of environmental issues provide early formative examples, cited by an adult who, though environmentally concerned himself, now effectively opposes much environmental legislation.

With regard to specific policies, Gerard favors nuclear power. He favors a large tax on gasoline, but would be concerned about equity effects of a tax on utility bills. He prefers incentives to drive less rather than fuel-efficiency regulations (he says that the latter imposes yet another burden on manufacturers and "a few thousand [automobile] fatalities a year"). Regarding our proposal for adaptation rather than prevention, he agrees that we should "not act preemptively without firm knowledge," but tempers this by saying that if climate change were definitively proven, "that would require some very firm response," and he would not be "cavalier" about conditions requiring moving farmers north.

In summary, Gerard's political conservatism easily coexists with his strong sense of environmentalism. He cites conservative principles to justify a noninterventionist position. If convinced that predictions of global warming were correct, he might favor prevention of climate change, rather than actively opposing it. However, Gerard has observed since elementary school that many predictions of environmental doom are overblown. He knows which members of Congress favor prevention of global warming and discards their evidence on the basis of their other political leanings. We should not forget that Gerard works for a member of Congress with a conservative constituency. All factors considered, he opposes action on global warming. His training as a lawyer leads him not to sift and evaluate competing scientific claims but to select and disseminate evidence supportive of his member's predilections, often in an entertaining and persuasive form.

Environmental sympathizers immersed in the political process blame individuals such as Gerard for lack of progress on prevention. One staffer we spoke with described Gerard and the representative he worked for as making a "vicious attack" on his arguments for preventing climate change. But we feel that one could just as well blame the scientific community and the media for not giving people such as Gerard a clearer sense of where the scientific consensus is.

In closing this section on congressional staff, we note that Gerard's view on climate change has prevailed in Congress as of this writing. Global warming has been used only as a secondary justification for multipurpose measures such as energy efficiency. The bills aimed directly at reducing greenhouse gas emissions, such as those drafted by Alvin and Mark, never made it to a floor vote.

Advocates

The two advocates in the next case studies are also influential, but their influence on government is achieved from the outside. Abby is a local environmental organizer and Bart is a Washington lobbyist for a national mining organization. Like the congressional staff, Abby and Bart both have some familiarity with global warming and have had to debate it in their advocacy activities, but it is only one of several environmental issues they must cover. They were picked because both are articulate and had a lot to say, and because they contrast with each other in the values given to justify the causes they advocate.

Abby (shop owner, environmentalist)
Abby devotes a substantial portion of her time to a local environmental organization concerned with both local and global environmental problems. She holds a bachelor's degree in sociology and supports herself by making hats, farming, and running an antique and secondhand shop. Boster, Kempton, and Hartley interviewed her in the millinery shop she was about to open in a New Jersey town. Although she claims no political party affiliation, she voted for liberal Democrats in recent national elections. Abby seemed sincere and dedicated to her environmental principles, and we found her personal philosophy unusually coherent among our informants. We discuss her philosophy and environmental values more extensively than those of the other case studies because she answered these questions in considerable detail.

Abby argues for environmental protection, saying that if we failed to maintain the environment, "we'd all be homeless, including the birds and the bees." She links environmental and social problems. It is her belief that the fulfillment of human and social needs depends on the health of the environment, the two are equally important. Making an argument frequently heard from environmentalists, Abby dismisses any conflict between protection of jobs and protection of the environment. She asserts instead that the health of both the economy and the environment is being sacrificed in the pursuit of military power. She argues that we should shift from an economy that exploits natural resources to make weapons to one that would use clean technology to provide for the needs of humans, plants, and animals. She also holds that the main cost of environmental protection is in the transition, and Americans can afford this cost by reducing waste (the example of waste she cited was wearing clothes only once or twice and then throwing them away).

Abby is heartened by recent political developments that signal an abandonment of the nuclear arms race. She hopes that the so-called peace dividend can be used to help solve social and environmental problems. Abby believes that the economy need not be sacrificed to protect the environment because many of the wastes that presently pollute the environment have potential uses, thus turning our problems into assets.

Abby views humans as arrogant, self-centered, and ignorant in their relationship with nature, but not deliberately cruel or evil. From her perspective, greed and ignorance lead to alienation from nature and hence to habitat destruction, and ultimately self-destruction, because people cannot foresee all of the harmful consequences of their exploitation of natural resources.

> Our society's become so alienated from the natural world, we're so bent on making a buck and blindly moving ahead that we just forget or we have blinders on and we walk all over it. It's not out of an evil heart that human beings act this way, it's just foolishness and ignorance.

Abby's ideal relationship between humans and nature is one of balance in which people weigh the costs of exploiting and disturbing the environment against the benefits they derive. "There needs to be a balance between what you do and what you get out of it and what the value is to what you do and the effect that you have." Although she acknowledges some forces of evil causing environmental destruction, the primary problem she sees is ignorance; the consequences of actions never get properly thought through and accounted for in decisions. She considers species extinction to be an especially bad example of the result of such thoughtless exploitation.

> We don't even know that they're gone. They came and went because we warmed up their lake or we screwed up their wetlands or their marsh. Things get out of balance because we're not understanding how everything we do has an effect.

Note how she extends the cultural model of "balance of nature" to a prescription for "balance" in human decision making. She says that we must maintain continual vigilance over the effects of our actions on the environment and modify our actions based on our understanding of those effects. Abby has the idealized view of traditional society and takes such societies as an exemplar for human treatment of nature.

I think there was a certain balance in the very beginning. If we want to try to get back to that, we should study the way native peoples have lived. There are some native peoples who didn't leave big scars where they lived, who were in balance. Almost always they had this great intuitive understanding of what they did. They never took too much, and they never wiped out a certain animal that they liked to eat because they always wanted some more there for the future.

A lack of spirituality contributes to what she views as humanity's ignorance of and alienation from nature. She wonders "whether we've become spiritually depleted as a society because we don't seem to think about the future." Though raised as a Catholic, she describes her spiritual outlook now as "sort of pantheistic."

I've experienced something bigger than what my own self contains. I think that people and living things somehow are united by some sort of force, or some sort of awareness or consciousness. And very few times in my life I feel like I've sort of tapped into that, and it's given me a glimpse of why life has any meaning at all. . . . Once, alone in the jungle in Mexico, I felt a unity with the natural world. I felt a presence of other intelligences, not mine. Something outside of myself that was not necessarily human, it was just some sort of consciousness. And I felt united and tapped into that. And sort of complacent about my physical surroundings.

This feeling of union with nature contributed to her strong feeling about the link between spiritual alienation from nature and exploitation of it.

It was very clear to me that the natural world is one that we really should not tamper with and that it's very difficult to get to that kind of feeling if everything is all screwed up, if everything is polluted and the air is foul and you're sick. I felt there had to be a certain level of health about the place and myself.

Abby maintains that the success of traditional societies in living in balance with nature is partly due to their being spiritually attuned to the natural world and the regular reenforcement of their beliefs through ritual. She sees in this a possible model for ourselves.

If our society had that [awareness] built into our lives through ritual, then I think it would be impossible for us to do what we're doing to the world because we just couldn't destroy what supports us, what gives us life.

Abby gave an extensive answer to our open-ended question about what she had heard about global warming. She mentioned fluorocarbons, nitrous oxide, carbon monoxide, and methane as source gases. As possible consequences she listed warming of the earth, rising of the oceans, drought, flood, increased salinity of rivers, water shortages, and devastation of ecosystems. This is an impressive list to produce off the top of her head—it is more complete, for example, than that of either of the congressional staff in the prior two case studies. (However, the list does seem to confuse carbon monoxide with carbon dioxide, and elsewhere in the interview she overextends the pollution model to greenhouse gases.) Abby's sources of information about climate change are quite diverse and include books on environmental science; newspapers (e.g., *New York Times, Philadelphia Enquirer,* local weeklies); magazines (e.g., *The Nation*); environmental publications (e.g., *Environmental Action, Greenpeace, Hazardous Waste News*); and radio (WBAR radio in New York, National Public Radio). She had also talked with environmental scientists and authors about their work.

Because global warming affects the planet as a whole, she favors international policy responses—in particular, adaptation of a worldwide agreement mandating substitutes for fluorocarbons, limiting the burning of coal and oil, and supporting the development of "clean technology." She opposes a shift to nuclear power ("just trading one poison for another"), preferring solar energy and conservation "on a really big scale."

Abby seems to consider the evidence for global warming convincing, although she acknowledges debate.

> I think there's a whole movement to say its not a problem, but I'm always suspicious of those movements. *Why?* . . . It always seems like those people with vested interests . . . producing things . . . are the loudest disbelievers.

Abby believes that adapting to climate changes ignores the basic problem. In her words, "The only way we're ever going to survive is to deal with the cause of the problem." She enthusiastically supports mandatory fuel economy standards, stating that auto companies already have the needed engine designs "hidden away in file cabinets." They are unwilling to use them, she says, because of their investment in outmoded technology and because "there are certain evil alliances between the petroleum industry and the car industry." She opposes carbon taxes because "you're taxing everybody, which means that people who can't afford to pay more are going to have to pay more, and that's not fair." Although increasing prices would cause some people to use less gasoline, "all that means is that the superconsumers get to really consume." In general, she believes that we can solve environmental problems by both developing clean technologies and adapting less wasteful ways to live.

In summary, Abby sees people's alienation from the environment as the primary problem. That alienation is destructive of both the environment itself and of the spiritual part of people themselves. Abby, a local activist in small-town New Jersey, has a knowledge of the science of problems like global warming that is arguably more complete than that of the three Washington-insiders described in this chapter. Yet she still makes extensive use of the cultural models described in this book: she invokes alienation from nature to understand society's environmental destructiveness, she extends "balance of nature" to become a prescription for balanced human decision making, she invokes native peoples as a guide to directions we should move toward, and her spiritual underpinnings provide strong and consistent motivation for her own action.

Bart (union lobbyist)

Bart is a lobbyist representing workers in the coal industry. He has a law degree and an undergraduate degree in history of government. He describes himself as a liberal Democrat and mentioned that he still believes in government command and control: "I'm one of the last ones in Washington." On the wall of his spacious Washington office is a map of coal deposits in the United States. This allows him to see at a glance which legislators should be most sympathetic to his union member's concerns. Bart spends a lot of time on "the Hill." In fact, Bart was one of the "coal boys" we watched file out of Alvin's office after the so-called beating up of Alvin about the proposed carbon tax. Alvin gave us Bart's business card, suggesting that we should interview Bart, too, to understand his position. During the interview with Kempton, Bart was easygoing, occasionally joking about his own biases derived from his job. He said he was "certainly more moderate" on environmental issues than most of his union members, noting that no one really extreme gets hired in Washington.

When asked whether protecting the environment is important, he says "sure," giving human health as the primary reason. By contrast, he volunteers that fishing does not seem to be such a strong reason for environmental protection. Asked about environmental values, he mentions that he participated in the first Earth Day. "I live in the suburbs, I like the outdoors," he says, continuing "everyone has a right to breathe clean air . . . and not be exposed to toxic waste or toxic emissions." He describes the relationship between humans and nature as "probably fairly well balanced." Bart does not believe humans have a responsibility to the environment.

> I consider it a responsibility to yourself, your community, your family. It's not the environment in abstract, I don't think of it as a being. . . . It's not an abstract good in itself. . . . Of course, that sort of gets into religious [pause] I don't usually think about that.

Whereas Alvin and Gerard felt environmental protection was in part for nature itself, Bart's reasons are clearly and unambiguously human oriented.

Like Gerard, Bart was ready for our question "*When people try to protect the environment, do you sometimes feel that other things don't get enough attention?*" Calling our question "a softball" (an easy question), he answered:

> A lot, yeah, absolutely. I mean, people are valiantly led into the trap of cleaning the environment, assuming there's really nothing to weigh in on the other side of the issue. . . . In Hollywood it seems like the big issue now is environment because there's no downside, there's no conflict.

As a specific example, Bart cited the debate on the Clean Air Act Amendments, which was winding down at that point.

> The science is a little shaky. . . . You remove X number of tons of sulphur from the air, what does that do to the fishing, the trout stream in Massachusetts. May have an effect, it may not. . . . Clearly sulphur dioxide is bad. . . . The question is [how much benefit will reductions give]. I think we'll never know. But we do know those mines are going to shut down, we do know those towns are going to lose tens of thousands of jobs, and we do know those families are really going to suffer; there'll be kids who aren't going to go to school, and kids who aren't going to have proper nutrition. It sounds like a real horror story.

Bart has heard about the greenhouse effect, both from media and from the environmental conferences he attends. Even though he has not yet studied it seriously ("Frankly, I'm not that familiar with the science of it"), he gives a reasonable brief summary of the causes and major effects.[8] Like Gerard and unlike Alvin, Bart sees the scientific community as divided.

I've heard Oppenheimer and other people. . . . I think people are kind of taking it [global warming] as fact whereas I gather in the scientific community, I've read some articles that say measuring temperature is not as exact a science as that . . . and the connection between CO_2 and temperature is not that obvious. Ten years ago they were talking about another ice age, and now they're talking about global warming, and there's not much change other than we've had a couple of warm years in the interim.

Given this uncertainty, Bart favors only modest steps. He would begin now with "more subtle" ways to reduce CO_2 emissions; he says "conservation is a good way to begin doing that." Bart would oppose a "major step," such as reducing CO_2 emissions 20 percent or 50 percent, which would mean that economic "harm is enormous and certain." Asked how he distinguished conservation from reducing CO_2 emissions 20 percent, he said "conservation in my own world means less coal, but I think it would not be very dramatic." Apparently Bart favors conservation because he presumes it would achieve less than 20 percent reduction in coal-derived energy demand.

Bart is also suspicious of the motivations of those promoting concern about global warming.

I mean, you can go the very cynical route which is it's the new issue to organize around, and there's money around; I don't know if that's the motivation of it. These people [environmental groups] are quick to jump on environmental horror stories just because they're convenient. People [the public] will believe it. . . . I think the problem is (this has been my class distinction here) the people who are going to argue for very severe changes are not the people who are going to bear the burden. They are the professional environmentalists, you know foundation people, people in the arts . . . to them it makes sense to do it because it's a good thing to do anyway. . . . But there's a point at which I think: consider other people.

This social-class consciousness is consistent with Bart's liberal Democratic politics. When asked who he felt had a completely wrong approach to dealing with the greenhouse effect, he mentioned Earth First!, which he had read about in a *New York Times Magazine* article. He notes their belief that the environment is an end in itself, and says he was "blown away" by their being quoted as saying that starvation and famine would be the natural order of things. Bart regards this position as lacking compassion and feels that Earth First! members are blind to the fact that their mandate to protect the environment "at all cost" entails human suffering.

> You can be as dramatic as you need to be to get through to these people "kids are gonna go without education, food, clothing, nutrition . . . high rates of divorce and suicide," I mean, that does happen in coal field communities. . . . I'd love to take them through Appalachia.

In this interview, our briefing on global warming was read in full, since Bart noted he was not yet totally familiar with the problem. After the briefing, Bart said the information was reasonably familiar, but that "I don't make the argument, [so] I don't keep track of it." At the point at which the interview briefing says that most scientists think there is at least a fifty-fifty chance that global warming will raise average temperatures, Bart interrupted to verify "Fifty-fifty?" His reaction at the end of the briefing was complex and interesting.

> Obviously everything you said stems from the assumption that the earth is getting warmer and the causes are [CO_2, CFCs, etc.]. . . . But, when I read opposing articles I blow them up in my mind, until they're actually as persuasive or more persuasive than the others.

Asked to summarize the opposing articles he had seen, Bart says:

> The change in temperature, when charted, doesn't match the increases in CO_2 and methane or whatever. And that even without [considering the correlation with] the causes, you can't show the temperature [increase] because of changes in measurement, urbanization, the major impact on temperature from the oceans. There's been very little done to measure changes in temperature of oceans or the air mass above the oceans. And there has not really been the depth of scientific inquiry necessary to say that at 50 percent {unintelligible} is a problem.

But Bart notes that he rarely sees these opposing views reported, except in "right-wing rags like the *Washington Times*." (Recall that this was also the source of Gerard's temperature graph.) Bart sees such articles because "People send me those articles 'cause they know I'm interested."

After mentioning that our briefing, like those he had heard at conferences, did not present the economic costs, he reflects on what if there really is a fifty-fifty chance of the effects of global warming as described in our briefing.

If this is in fact what's going to happen, I think this takes precedence over the economic impact. Lord, this will have an economic impact of its own which needs to be taken into account; I think then you have to start working right now on the transition. How do we want to cut 50 percent of coal and natural gas, what do we do about those people and communities that depend on their production. Or . . . my preferred way of dealing with it is a Los Alamos type of study of technologies to control emissions. We've developed the technology for sulphur dioxide . . . we've apparently not done it for carbon dioxide.

Although he had remained unconvinced by several presentations on global warming, he was clearly moved by our briefing. This may be in part because he saw us as a more objective source than the environmental activists at the conferences. (Anticipating such perceptions and our concomitant responsibility as interviewers, we had worked hard for balance and verified scientific support for each point in our briefing.) We find it amusing that the same briefing that produced an angry tirade from Alvin for downplaying the effects and being insufficient to convince the "coal boys" did in fact produce from "coal boy" Bart a concerned reaction.

Of course, our briefing was not sufficient to change Bart's mind for long. He continued to consider the evidence predicting climate change "not very persuasive." Furthermore, if he were convinced, his fallback position is to continue burning coal and remove the CO_2. Bart raised the possibility of pollution controls for CO_2 emissions at a conference on global warming and was told that it was impractical, that he should focus on how to eliminate burning of fuels. We assume that this advice was due to the technical difficulty of removing CO2. Bart interpreted this advice more personally, inferring that the speaker was not concerned with the effects of ceasing production, whereas he thought it should be the "overriding concern."

Once you've decided this is the impact on the environment, your next line of inquiry could be: How do we preserve mankind while we make these changes? *When you say "preserve mankind," you mean preserve our ability to produce power and food and everything else?* Right, right.

Note that he is using "preserve mankind" very differently from how an environmentalist would use the phrase.

In sum, Bart bases his moral judgments on human needs and values, not seeing nature as an entity on its own. He is one of the clearest and most unambiguous respondents on this. Consistent with both his liberal Democratic politics and his constituency, he is concerned about the plight of coal miners thrown out of work by environmental regulations. He perceives a lack of concern from a range of environmental advocates, ranging from wealthy Hollywood actors to hard-hearted environmental radicals, and suspects the motives of those raising concern about global warming. He sees no consensus from the scientific community, but when given a thorough briefing from what he considers a credible source (us), he temporarily switches into thinking seriously about dealing with the problem.

Conclusions

We presented these four cases in order to show how beliefs and values fit together as a coherent whole within individuals, to give a more integrated view of environmentalism (leading into the next chapter), and to show how information is selected in order to fit with one's predilections and institutional needs. We picked influential specialists, both because specialists are in an interesting intermediate position between scientists and the public and because the type of specialists described here have a disproportionately large effect on policies and laws.

Stereotypes about those who oppose environmental regulations explain little here. Bart is a liberal Democrat, thoughtful, self-reflective, and concerned about other people. His compassion ends at the boundary of humanity—it just makes no sense to him that the environment is a separate entity deserving of consideration or rights. Gerard breaks the antienvironmentalist stereotype even more. Gerard likes the outdoors and values environmental protection not only for human utility but for the environment itself. In earlier chapters, we quote his moving arguments for protection of species and his cautions against major environmental disturbances. So how can we explain Gerard's active opposition to environmental legislation?

Gerard believes that many environmental warnings are overblown and based on unreliable computer models, that advocates are often biased or have hidden agendas, and that many environmental protection measures are not worth the cost. A key statement from Gerard is that "politics drives the environmental issues in a really fundamental way." He notes biases leading to an overemphasis on environmental issues and can cite multiple examples as evidence, ranging from his elementary school, to city council resolutions in his district, to the Clean Air Act. If people were convinced by his position, Gerard feels, voters would still have a "strong environmental concern" but "a more targeted and intelligent one." If we had extracted some of Gerard's separate quotations without a full case study, he might seem to contradict himself. The case studies show that the contradictions are not in people like Gerard, but in simplified stereotypes we hold about the advocates and opponents of environmental policies.

We found multiple processes that select and filter information so as to buttress ones own (or one's organizations') ideology and interests. Alvin's anger at our "trivial" list of global warming effects is one example. Gerard humorously exaggerates problems with environmental standard setting, and selectively cites data for a "Dear Colleague" letter to the entire house. Bart apologetically turns to a "conservative rag" because it seems to be the only source of alternative views on global warming.

Both of the congressional staff evaluate scientific evidence based on the interests, political leanings, and prior statements of the presenter of the evidence. This evaluation procedure is used for other members of congress as well as scientists. Similarly, these advocates and political actors also make much reference to characteristics of people taking contrary positions, saying they are ignorant, lacking human compassion, motivated by hidden agendas, or influenced by political bias.

Among these specialists, we find fewer of the confusions noted in the chapter on models of weather and the atmosphere. For example, only one of these four specialists inappropriately applied a pollution model to global warming, and only one combined global warming and ozone depletion (some staff reported that members of congress also combined global warming and ozone depletion). As noted for the specialists in the previous chapter, these specialists differed greatly in the depth of knowledge they had of different aspects of the problem.

One important difference that emerged among these four cases was their assessment of consensus among scientists. Alvin and Abby saw widespread agreement, considering those scientists who challenge predictions of global warming to be primarily from affected industries. Gerard and Bart saw a complete spread of opinion; as Gerard said, "there's no mainstream [of expert opinion], there's no fringes, there are just people all over the lot."

Global warming is a complex, difficult to evaluate issue. We should not be surprised if the political process handles it imperfectly, with a greater diversity of views among the politically involved than among scientists themselves. Is the process of science advising to the legislature flawed? There are more systematic means for transferring information from scientific studies to political decisions. The U.S. legislature already has the Office of Technology Assessment, which had not yet produced a report on global warming at the time of our congressional interviews. There are other examples abroad, such as the Intergovernmental Panel on Climate Change (e.g. IPCC 1992), in which scientists from many nations meet and reach consensus on what findings can be agreed upon. In Germany, global warming was investigated by a special committee which made up of half scientists, and half government officials, meeting weekly over several years (see their report, Bundestag 1989, and observations on the process, in Kempton and Craig 1993 and Craig, Glasser, and Kempton 1993). Based on our case studies, it would seem that more systematic, institutionally supported access to scientific and technical judgments surely would have worked more satisfactorily than the current systems of dissemination.

8

Patterns of Agreement and Disagreement

We now broaden our perspective from topics or individuals to the overall patterns of agreement and disagreement in the whole sample. In chapters 3 through 6, the tables that presented our survey results showed how broadly environmental beliefs and values are shared across the five groups responding to the survey. In this chapter, we apply two analytical methods to explore more systematically the nature and extent of this sharing. First, we use graphs to examine the degree to which people and groups answered our survey questions similarly. This tells us who agrees in their beliefs and values without specifying what they agree or disagree on. Second, we tabulate the answers to the survey statements that received strong agreement or disagreement in order to generalize about the content of what Americans share and what they do not.

Patterns of Agreement

To examine patterns of agreement, we use analytic methods developed and refined in anthropological studies of intracultural variation (the pattern of agreement and disagreement among individuals in a culture). These methods set guidelines, described in detail in appendix A, for determining whether or not there is a *cultural consensus,* that is, a generally agreed upon set of beliefs and values. Other possible results from these analytic methods are multiple consensuses, each held by different subsets of individuals, or no consensus at all.

We introduce our methods for analyzing patterns of agreement with an example of an earlier anthropological study (Boster and Weller 1990).

The study compared two cultural groups, corn farmers in Tlaxcala, Mexico, and college students in Pennsylvania. Members of each group were asked to classify a series of foods as being either hot or cold. A graph resulting from the analysis is shown in figure 8.1. Each of the points on the graph signifies an individual, with the point's position indicating how similar the individual's answers are to answers of the other people surveyed. The individual's agreement to the overall group is indicated on the horizontal axis of the graph. Individuals near 1.0 indicate strong agreement, those around 0 indicate no agreement, and negative values indicate an individual holding views opposed to the remainder of the group. The vertical axis measures remaining differentiation not already captured by the horizontal axis. Negative or positive values on the vertical axis do not have any direct interpretation.

The conclusions that can be drawn from figure 8.1 illustrate how graphs such as this are interpreted. First, the spread of points above zero along the horizontal axis indicates that respondents do not radically

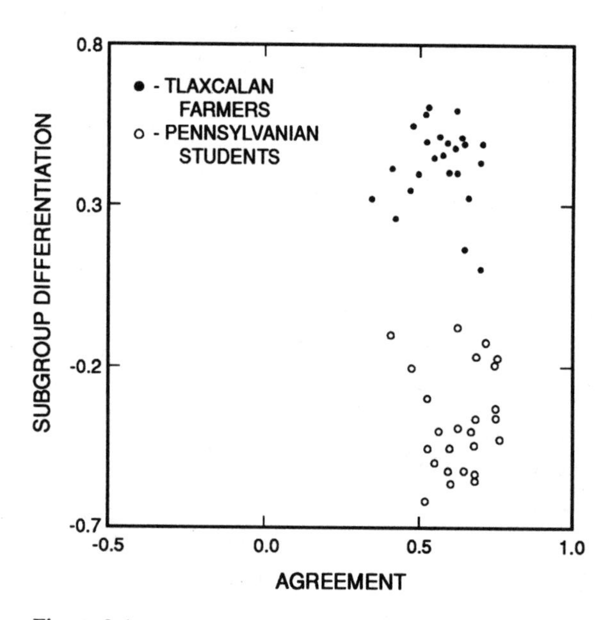

Figure 8.1
Example study: Two separate consensuses found in a cross-cultural study of food classification

disagree in their classification of foods as hot and cold, even though the respondents come from two different societies. However, along the vertical axis, the clustering together of points for Tlaxcalan farmers, and the separate cluster of points for Pennsylvania students, indicates that there was not a single consensus. That is, despite some internal variation, each of these groups has its own distinct cultural system for classifying hot and cold foods. (The substantive explanation is that Tlaxcalans classify food more on the basis of how one feels after eating the food, while the Pennsylvanians classify more on the basis of ambient temperature when served.)

Studies applying this analytic method within a single culture may find one consensus or multiple ones. For example, Boster and Johnson (1989) compared recreational and novice fishermen in judgments of similarity among fish. They found two separate consensuses, each a coherent and consistent set held by one subgroup. Conversely, Weller, Romney, and Orr (1986) found a single cultural consensus on appropriate childhood discipline, despite expectations of subcultural differences.

We expected a distribution of individuals something like figure 8.1 in our environmental survey. Previous work by Dunlap and Van Liere, Milbrath, and others on the "new environmental paradigm" versus the "dominant social paradigm" suggest a split between two views of the environment, each a consensus view of a subgroup, like the graph of figure 8.1.

To explore the patterns of agreement in our survey of environmental beliefs and values, we present a series of graphs similar to figure 8.1. We first consider the results obtained by giving our survey to the individuals whom we had already interviewed. We then examine the results obtained for sawmill workers, dry cleaners, laypeople, Sierra Club members, and Earth First! members.

Using the criteria of the cultural consensus model (Romney, Weller, and Batchelder 1986), we found that there was enough agreement on beliefs and values to treat almost all survey respondents as members of a single cultural group, with a shared set of beliefs and values. (The criteria and their application to our data are described in Appendix A.) A cultural consensus does not mean universal agreement, for some statements are more widely agreed upon, and some individuals share more of the consensus view. Rather, it is shared in the way that, for example, geographical knowledge in the United States is shared; there is only one cultural system that individuals share to varying degrees.

Figure 8.2 shows agreement among the people who took part in both our semistructured interview and our survey. Of these twenty-six individuals, half are clustered in the upper right-hand corner of the graph. The rest are distributed in a roughly diagonal pattern from the cluster to the lower left-hand corner.

The axes of figure 8.2 and subsequent graphs may be interpreted as follows. The horizontal axis measures how close each individual is to the overall consensus among all respondents. That is, in figure 8.2, the respondents who agreed most with other respondents are at the right of the figure. The vertical axis reflects the greatest remaining differences among respondents. On inspection, those with the most proenvironmental responses are found at the top, and those with the least pro-environmental responses are at the bottom. Therefore, we interpret this axis as a component of environmentalism not captured by the horizontal axis.

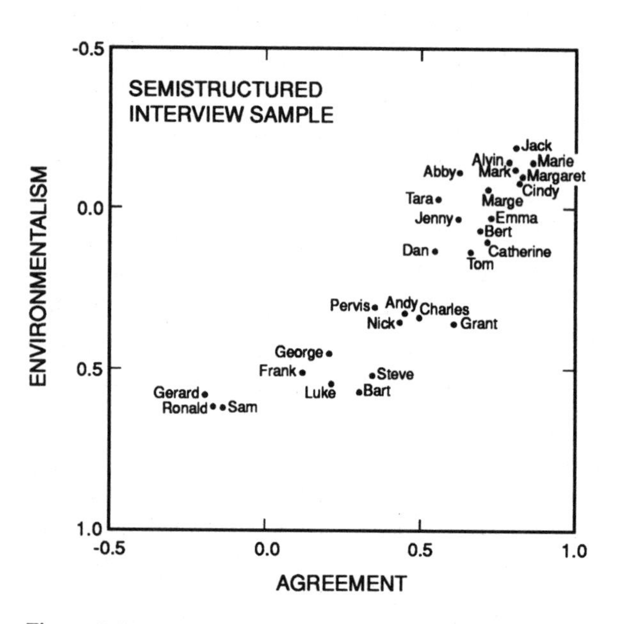

Figure 8.2
Informants from semistructured interviews, located on agreement graph

The labeling of the horizontal axis as "agreement" is not based on interpretation but is a direct result of the analytical method. Since the agreement is on questions concerning the environment, the axis labeled "agreement" must reflect environmentalism as well. Of the two axes, the horizontal axis is the more important, capturing five times as much of the variation in the data as the vertical axis.

We label the data points in figure 8.2 with the individuals' names, in order to track the informants discussed in our case studies and thereby cross-check the analysis producing the graph. Informants who gave the most proenvironmental responses in their interviews are in the upper right cluster. They include the environmentalist Abby and congressional staff member Alvin, whose case studies were presented in the preceding chapter. In contrast, individuals who gave the least environmental responses in the interviews are farther to the left and lower on the graph. They include Bart, the representative of mining interests, and Gerard, the conservative congressional staff member.

Consider how various aspects of the individuals in the case studies are reflected by their position in figure 8.2. Our case studies make clear that Alvin and Abby differ in many ways—for example, Alvin stressed the utilitarian value of the environment whereas Abby felt a more spiritual motivation for her environmental activism. Nevertheless, they are clustered close together in the graph. This means that, despite different values underlying their environmentalism, they answered most statements on the survey similarly to each other. On the other hand, even though Gerard valued species preservation and held the major cultural models of the environment, he opposed a number of regulations and thought many environmental concerns exaggerated. His position on the horizontal axis, just below zero rather than a large negative value, shows that he does not share much of the environmental view, but he is not directly opposed to it either.

Going beyond the cases of chapter 7, we see that in the lower and middle left, close to Gerard and Bart, are two laypeople picked because they were believed to oppose environmental laws (Ronald and Frank) and some of the members of groups expected to be less sympathetic to the environment (congressional staff opposing environmental regulations, workers in coal and automotive industries). All the laypeople picked at random from the public, and many of those from industries, are in the upper right portion of the figure, near Alvin and Abby.

When we graph the patterns of agreement in our five surveyed groups, an interesting pattern emerges (figures 8.3–8.7). The order in which we present the resulting graphs reflects the expected order from less to more environmentally concerned.[1] The sequence goes from laid-off sawmill workers (figure 8.3), to Los Angeles dry cleaners (figure 8.4), to laypeople in California (figure 8.5), to Sierra Club members (figure 8.6), and finally to Earth First! members (figure 8.7). We find that the data points for each successive group are further to the right, higher, and more tightly clustered. The shifting locations of data points with each graph indicate three simultaneous changes in the patterns of agreement among the groups: (1) rising levels of agreement with the consensus (points further right on the graph), (2) increasingly proenvironmental responses (points higher and to the right on the graph), and (3) greater within-group homogeneity (points closer together). Thus, the sawmill workers are spread out across the graph, representing the whole range of views. In

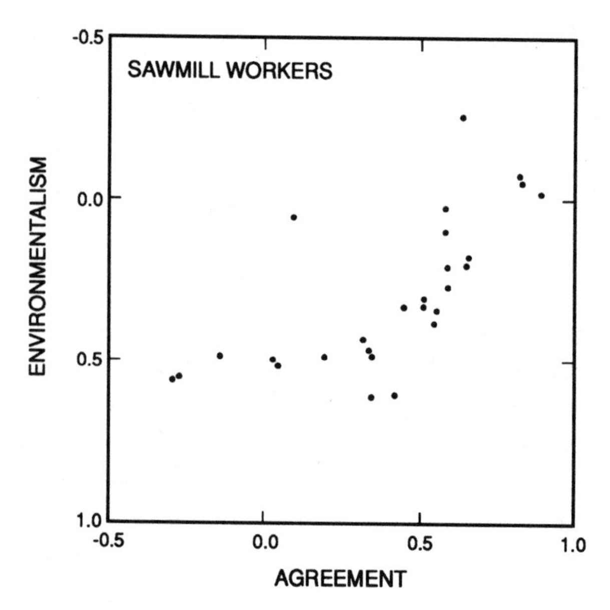

Figure 8.3
Sawmill workers on agreement graph

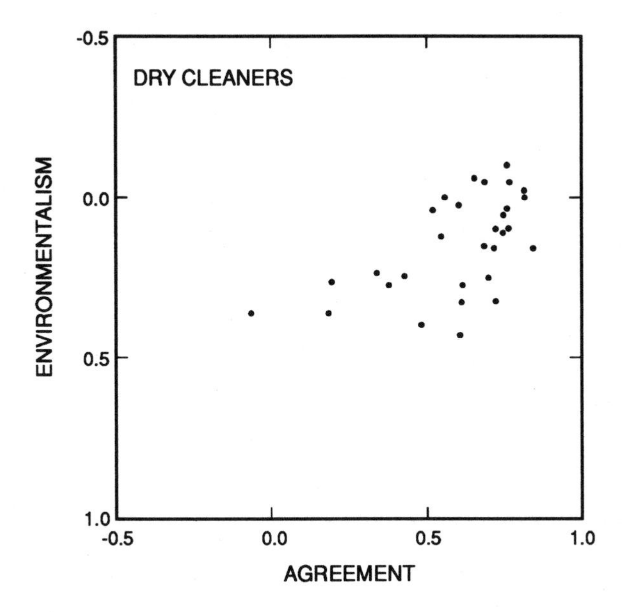

Figure 8.4
Dry cleaning managers on agreement graph

contrast, Earth First! members all share a strong environmental consensus. The dry cleaners, lay public, and Sierra Club members are intermediate between these two extremes.

The clustered distributions of Earth First! and Sierra Club members are not surprising and neither are the positions of their clusters relative to one another. We expected each of them to exhibit an environmental consensus, with Earth First! members positioned further toward the environmental end than the more mainstream Sierra Club members. What is surprising about the graphs is the distributions found for the sawmill workers and dry cleaners. These groups were selected for study because we expected that they would be antienvironmental as a result of their livelihoods having been hurt by environmental laws. However, in the graphs of the sawmill workers, and to a lesser extent the dry cleaners, the data points are widely scattered. While several sawmill workers and some dry cleaners did give antienvironmental responses, ten of the twenty-seven sawmill workers and twenty-one of the thirty dry cleaners gave responses that would place them within the cluster of

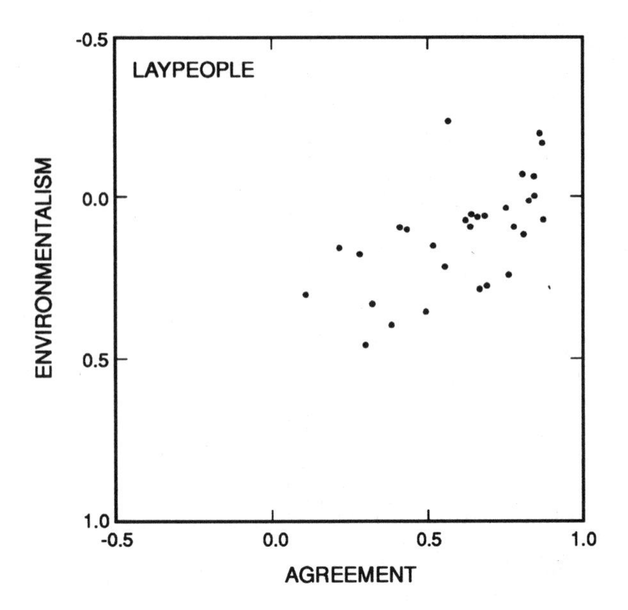

Figure 8.5
Sample of lay public on agreement graph

points of the Sierra Club members. Four of the sawmill workers gave responses that would place them in the cluster of points of Earth First!. Thus, although members of these groups were chosen for this study because we expected them to be antienvironmentalist, many of them agreed with the proenvironmentalist consensus. This level of agreement was even stronger among the laypeople whom we studied, two-thirds of whom gave answers that fell within the cluster of Sierra Club members.

One conclusion that can be drawn from figures 8.2 through 8.7 is that, despite a range of variation in environmental views, there is an environmental consensus among a majority of the people whom we interviewed. This majority includes not just environmentalists but almost all of the members of the general public and many of those from groups expected to be antienvironmental as well. In other words, the views of the general public on environmental issues match most closely those of self-identified environmentalists. There is no cluster of respondents strongly negative on the "agreement" scale, as there would be if there were a clear antienvironmental position. Furthermore, there is no cluster

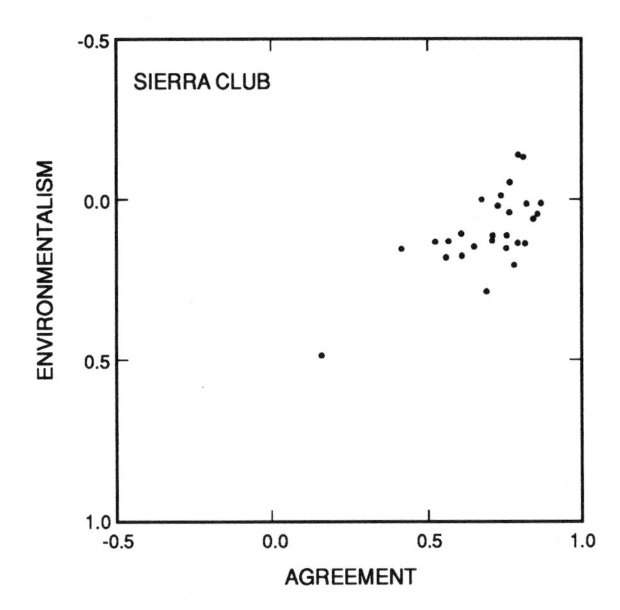

Figure 8.6
Sierra Club members on agreement graph

of respondents apart from those at the environmentalist end of the graph, indicating that no subset of respondents agrees on a consistent alternative to the environmentalists' survey responses. The few who do not agree with the cultural consensus on environmentalism do not share among themselves a coherent and consistent alternative.

Expected versus Actual Patterns of Agreement

Our findings challenge both prior research on this subject and the ways environmental opinion is portrayed in the media and in popular stereotypes. In previous writings by Henderson (1976) and Harman (1977), and in previous research by Dunlap and Van Liere (1978), Milbrath (1984), and others, two opposing paradigms are postulated. These scholars contrast the existing establishment position, which they call the "dominant social paradigm" (DSP) with an emerging "new environmental paradigm" (NEP). As reviewed briefly in chapter 1, these authors use

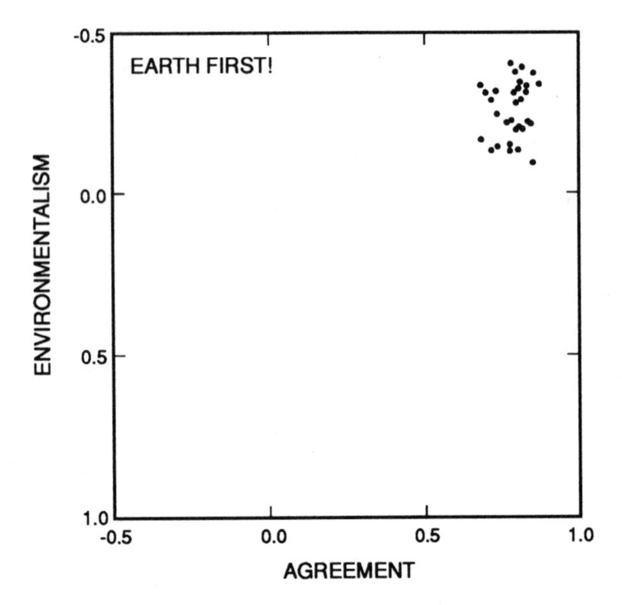

Figure 8.7
Earth First! members on agreement graph

"paradigms" in the sense of Kuhn (1962) to denote comprehensive ways of viewing the world, each mutually exclusive.

Milbrath's (1984) study illustrates the conclusions of this previous work. In formulating the questions for his survey, Milbrath, like earlier scholars, relied heavily on the writings of environmental activists. The resulting questionnaire asked respondents about their views on the environment as well as on other possibly related issues such as the role of government planning versus market decisions, participatory politics versus determination by experts, and cooperation versus competition in solving problems.

Milbrath administered the survey to a sample of individuals representing the U.S. general public, environmentalists, and leaders from government, business, and labor. Milbrath interpreted their responses to his survey as confirming the existence of the two paradigms and as showing them to be differentially distributed among people. The dominant social paradigm was held more often by leaders from government, business, and labor groups, and the new environmental paradigm was held more often by environmentalists. The general public leaned slightly toward the environmentalist position, but were near the center: "the mass of people in the United States . . . partake of the beliefs of both paradigms and thus can be plotted near the center of the space" (1984, 23). Figure 8.8 presents a schematic summary of Milbrath's findings (1984, 24).[2]

Milbrath's results seem to empirically confirm what may seem obvious to many: that environmental issues often involve opposing positions that are held by different groups. This dialectical position reflects in many ways the manner in which our society approaches many issues, as seen in news reports quoting representatives of both sides of an issue, in political debates setting pro against con, and in legal cases argued by opposing advocates. Our society's processes for dealing with environmental issues airs conflicts rather than finding common ground.

We expected that our survey results would reveal patterns similar to those documented by Milbrath (1984). Clearly they did not. Figure 8.9 presents a summary of our findings using the same schematic format used to summarize Milbrath's findings (figure 8.8). A comparison of the two figures illuminates the major differences between the two sets of survey results. First, in our figure, the public is not near the center but is shifted well over to the environmentalist side. Second, the degree of separation between the public and environmentalists in our figure depends on the environmental group being considered. Earth First! members are outside the range of (although very close to) the general public's environmentalism, whereas Sierra Club members are entirely within the public's range. A final difference between our survey results and those of Milbrath is that even though our survey sampled two groups that had been harmed by environmental regulations, we found no coherent alternative to the environmental paradigm. This finding is represented by separate dots rather than a circle on the antienvironmentalist side of figure 8.8.

Why do our results differ from those of Milbrath? We believe there are two reasons, which lead us to conclude that our study's results better represent the current spectrum of American environmental views. First, our survey questions were derived from laypeople's statements about the environment in semistructured interviews. In contrast, Milbrath's ques-

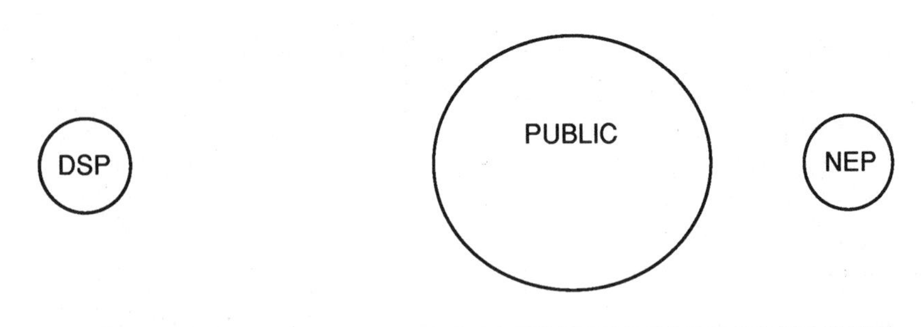

Figure 8.8
Schematic summary of Milbrath's findings. The new environmental paradigm (NEP) is opposed to the dominant social paradigm (DSP).

tions drew heavily from the writings of environmental thinkers and activists, whose concerns often include not just environmental issues but broader social change issues as well, such as the logistical problems of establishing voluntary organizations, whether the country should emphasize cooperation or competition in solving problems, and whether people should be judged on their personal qualities or their achievements. Consequently, Milbrath's questions included many that we believe are unrelated to lay environmentalism. Most important, his diagram was explained in terms of answers to three key questions: whether the environment is a small or large problem, whether environmental problems are better solved by better technology or "basic change in society," and whether there are limits to society's growth (1984, 43–49). We feel that answers to the second and third questions could well be based primarily on issues unrelated to the environment.[3] The diversity of survey questions used by Milbrath and others to study environmental paradigms has recently been questioned by Olsen, Lodwick, and Dunlap (1992, 88–89). In our survey, we included very few nonenvironmental questions, because such statements were not volunteered by our informants when they were discussing environmental issues.

The other reason we believe our survey results provide a better representation of American environmental views than Milbrath's is based on our analytical procedure. Milbrath derives his diagram from both interpretation of survey results and theoretical considerations. Thus, his

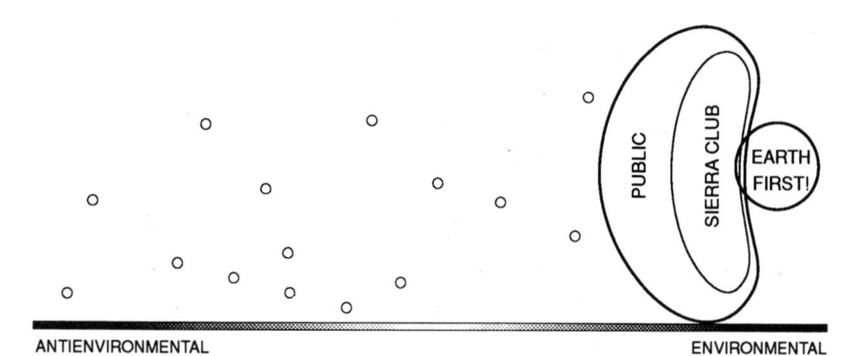

Figure 8.9
Schematic summary of our findings, derived from figures 8.2 through 8.7.

diagram (our figure 8.8) is not based on a formal analytic procedure. In contrast, we used a method specifically designed to determine how similar the views of different individuals and groups are. The resulting graphs (figures 8.2–8.7) in turn provide a basis for generating the schematic diagram (figure 8.9) summarizing our results.

One other possible explanation of the differences between Milbrath's and our findings is that in the decade since Milbrath's data were collected there has been a change toward greater environmentalism by the general public and less coherence among those questioning environmentalism. It is also possible that in the same period environmentalism has become decoupled from the social change agenda inherited from early environmental activism in the sixties. While we acknowledge that these changes may in fact have occurred, we would argue that their effect on our survey results would be smaller than the substantial differences in types of questions and methods for analysis we outline above.

In sum, we believe that the differences between Milbrath's and our findings can best be explained as a consequence of the different methods we used to generate our survey questions and to analyze our data. If the goal is to provide a description of how environmental disputes play out in the media and the political arena, then Milbrath's diagram (figure 8.8) may be the most accurate. However, if the goal is to understand whether or not environmental paradigms exist among the public and, if so, which citizens share them, then we believe that our methods provide the more accurate picture.

Consensus Beliefs and Values

We turn now to the content of the cultural consensus: on what statements do most people agree?

To investigate the content of the environmental consensus, we determined which of the 149 statements in our survey were widely accepted or rejected by Earth First! members as well as sawmill workers. We look for agreement across these two most divergent groups as a simple way to identify the beliefs and values most widely shared in the society. We compare first the statements both groups accept, then the statements both groups reject.

We first examine the statements for which two-thirds or more of both groups agreed. Forty-six statements met this criterion, a representative sample of which are in table 8.1. The two-thirds cutoff has no special theoretical significance, but we judge two-thirds agreement of each of these two extreme groups to indicate a broad consensus. (The interested reader can scan appendix C for the complete set of statements meeting this or any other criterion.)

The statements receiving two-thirds agreement by both Earth First! members and sawmill workers roughly correspond to the cultural models discussed in earlier chapters. For example, both groups agree that there is a need to protect the environment because humans depend on it (statements 7, 65, 108). Both groups share the model of interdependencies in nature and the consequent nonintervention model (48, 57, 109, 126); both are skeptical about corporations or the profit motivation (34, 55, 89, 144). They expressed the value of environmental protection for one's descendants (24, 27, 111), and, more surprisingly, concurred with a value of nature for its own sake (51) or because it is God's creation (58). They also agree on the desirability of reducing consumption (22, 141, 147) and on the prudence of prevention (31).

At least two-thirds of the Earth First! members and the sawmill workers reject twenty-three statements, a representative set of which are in table 8.2. Both groups reject claims that current environmental protection is excessive (29, 74, 131, 132), that nature's value is only in serving humans (75, 80), that government, not individuals, is responsible for solving global warming (11, 68), that evidence of global warming is insufficient to act (90, 130), and that nuclear power should be adopted more widely because it does not pollute (128). Some of these are components of the cultural models discussed earlier and some are separate beliefs or values.

We note that shared agreement or disagreement is easier for the statements in tables 8.1 and 8.2 because the majority of those statements do not require choices between conflicting values. (About thirteen do require choices between conflicting values, and nevertheless the two groups agree.[4]) Overall, high agreement between these supposedly diametrically opposed groups illustrates that there are quite a few widely shared environmental beliefs and values, supporting the earlier graphical analysis of patterns of agreement.

Points of Dissension

Although our survey has revealed significant agreement in the environmental views of our surveyed groups, it also shows some significant differences. To document the content of these differences, we again compared the responses of the two most divergent groups, Earth First! members and sawmill workers. We examined those questions for which two-thirds or more of the Earth First! members agreed and two-thirds or more of the sawmill workers disagreed, or the converse. All of these statements are listed in table 8.3.

Many of the statements that Earth First! members and sawmill workers disagreed upon involved a choice between values. Trade-offs included the environment versus feeding one's family (63), vegetarianism (99), the environment versus social justice (92), and the rights of other species versus those of humans (78, 84). In addition to trade-offs, the two groups disagreed on whether reducing pollution prevented climate change (116); we would explain this by saying that sawmill workers applied the pollution model incorrectly, Earth First! did not. The groups disagreed on several statements that could be argued to be either an assertion of opinion or a test of incorrect use of models (45, 37, 92). They disagreed

Table 8.1
Subset of survey statements accepted by at least two-thirds of Earth First! members and sawmill workers

Earth First!	Sawmill workers	Survey statements
93	74	1 The weather has been more variable and unpredictable recently around here.
68	74	2 We are not going to be able to totally prevent global climate change, so adaptation has to be part of the solution.
100	70	5 Change of a few degrees in average temperature can make a huge difference in whether things can grow or not; global climate change would mean there would be enormous disruptions of agriculture.
94	100	7 People have a right to clean air and clean water.
77	93	22 In World War II, people gladly rationed for the war effort. People would willingly do it again to save the environment if the need were great enough.
100	74	24 We have to protect the environment for our children, and for our grandchildren, even if it means reducing our standard of living today.
100	96	27 We have a moral duty to leave the earth in as good or better shape than we found it.
93	85	31 Preventing global climate change is better than waiting to see what happens. It's more costly to fix problems than it is to prevent them in the first place.
83	89	34 They could easily design a much more fuel efficient car, but the big automobile companies conspire with the big oil companies to keep that from happening.
86	67	48 We don't have the right to play God by manipulating nature.
97	82	51 Our obligation to preserve nature isn't just a responsibility to other people but to the environment itself.
97	85	55 The reason politicians break their promises to the people to clean up our environment is the power of industry lobbies.

Table 8.1 (continued)

Earth First!	Sawmill workers	Survey statements
100	82	57 Global climate change would disturb the whole chain of life.
76	78	58 Because God created the natural world, it is wrong to abuse it.
100	78	65 We spend billions of dollars trying to find the cure for cancer and yet we're not doing enough to clean up the environment which causes the cancer.
97	78	89 Unfortunately a lot of companies wouldn't do anything to protect the environment unless they were forced to by law.
84	93	108 If global climate change would cause things like a million people drowning in Bangladesh, we can't let that happen. Even if it doesn't affect us here, that would be totally immoral.
94	85	109 Nature may be resilient, but it can only absorb so much damage.
97	85	111 Working to try to prevent environmental damage for the future is really part of being a good parent.
94	67	126 Global climate change would be bad even if it didn't cause humans any harm, because it is not a natural change.
100	70	141 We should return to more traditional values and a less materialistic way of life to help the environment.
90	67	144 Government restrictions on the use of private property are necessary in order to insure that the land will not be permanently harmed.
90	67	147 Americans are going to have to drastically reduce their level of consumption over the next few years.

Note: The first two columns indicate the percentage agreeing with each statement.

Table 8.2
Subset of survey statements rejected by at least two-thirds of Earth First! members and sawmill workers

Earth First!	Sawmill workers	Survey statements
7	15	11 Because global climate change is a world issue and not an individual one, it's the government's responsibility to do something about it, not mine.
0	33	29 The environment may have been abused, but it has tremendous recuperative powers. The radical measures being taken to protect the environment are not necessary and will cause too much economic harm.
29	19	68 The public doesn't need to know all the details on policies to deal with global climate change, Congress just needs to pass the right laws.
14	26	74 The environment probably doesn't need as much protection as we imagine.
0	11	75 People's only responsibility to nature is to make it serve their own best interests.
0	31	80 Plants and animals are there to serve humans. They don't have any rights in themselves.
0	22	90 Scientists are just speculating about global climate change. We shouldn't take action until they have proof.
3	26	128 Nuclear power doesn't pollute as much as other power sources. We should expand our use of it for the sake of the environment.
32	30	129 We should not force the auto companies to make cars with higher gas mileage. Instead we should discourage people's excessive use of their cars.
0	30	130 We shouldn't take action on global climate change based just on computer models and scientist's predictions. We should wait until we can measure an actual temperature change and know it's real.

Table 8.2 (continued)

Earth First!	Sawmill workers	Survey statements
0	0	131 Humans can't live without creating waste. We should use the old-fashioned methods—find some place out in the woods where it doesn't bother anybody, and just dump it.
0	15	132 We shouldn't be too worried about environmental damage. Technology is developing so fast that, in the future, people will be able to repair most of the environmental damage that has been done.

Note: The first two columns indicate the percentage agreeing with each statement.

on some statements that were simply matters of preference or belief (125, 121). Predictably, they also disagreed on statements about environmentalists: whether environmentalists were naive (96) and whether environmentalists' views would change if their own jobs were threatened (123).

Variation and Environmental Groups

We selected our five survey groups in order to sample diversity in American culture, not because we wanted to understand these groups in and of themselves. Nevertheless, our general findings are enriched by considering in greater detail the two environmental groups, Earth First! and the Sierra Club.

The Earth First! members have been surprising in that, for a group widely considered radical, each of our analyses has shown them to be not very different from the rest of our respondents. In the graphs of the patterns of agreement, they were found to share the same cultural consensus of other Americans, differing only in answering more consistently. In the analysis of the content of the cultural consensus, we find that they agreed on many statements with all other groups, including the sawmill workers. In the analysis of points of dissension, Earth First! members

Table 8.3
All survey statements for which at least two-thirds of Earth First! agreed and less than one-third of sawmill workers did, or the converse

Earth First!	Sawmill workers	Survey statements
100	30	42 Energy conservation will work better if we price energy correctly through higher fuel taxes to make efficient energy use in people's own interest.
73	22	127 We shouldn't resort to nuclear power, even if the industry can make it completely safe, take care of wastes, and prevent it from going into bombs.
13	70	63 My first duty is to feed my family. The environment and anything else has to come after that.
80	7	78 If any species has to become extinct as a result of human activities, it should be the human species.
90	22	84 I would rather see a few humans suffer or even be killed than to see human environmental damage cause an entire species to go extinct.
23	67	37 We don't have to reduce our standard of living to solve global climate change or other environmental problems.
13	67	125 If we had the technology to change climate for the better, to improve the human condition, we should do so.
90	22	99 We should become vegetarians to reduce our environmental impact.
24	74	45 I don't believe in preservation of species in the way some environmentalists do. In nature, evolution includes extinction.
21	67	116 Reducing pollution is a more effective way to prevent global climate change than energy conservation.
17	81	92 A fuel tax would be an unfair way to reduce fuel consumption because some people are forced to use more fuel than others by their business or personal needs.

Table 8.3 (continued)

Earth First!	Sawmill workers	Survey statements
77	33	121 Nature is inherently beautiful. When we see ugliness in the environment, it's caused by humans.
25	70	96 Some of the people who are the most passionate about global climate change are pretty naive about the facts. The ones who know what they're talking about tend to be cautious about making predictions.
23	78	123 Environmentalists would not be so gung-ho if it was their jobs that were threatened.

Note: The first two columns indicate the percentage agreeing with each statement.

and sawmill workers did disagree often, but this was usually dissension over the relative ranking of environmental with other values, rather than disagreement on the values themselves.

Our conclusion contrasts with the usual picture of Earth First! in the media, which portrays them as holding a form of environmentalism quite different from other Americans. Our analysis suggests that they have more or less the same beliefs and values as other Americans. This account differs from media stereotypes, but it dovetails with descriptions from scholars who have done in-depth ethnographic research on Earth First!. For example, Bron Taylor (1991; 1993) describes them as being radical in their willingness to act, such as by blocking a woodcutter's axe by placing their hands against a threatened tree. But Taylor concludes that they are not different from average Americans in what they believe and value. In short, it is not that Earth First! members have unusual values but that they are willing to make greater personal sacrifices for their environmental values.

Although Earth First! members and sawmill workers differ in their views on some environmental issues, the disagreements were not as numerous as we would have expected. This cautionary lesson—that we should not make a priori assumptions about which groups will disagree and how extreme their views are—is also reenforced by the responses of members of the Sierra Club local chapter we sampled. Sierra Club members were expected to represent moderate environmentalism. To our surprise, however, some of the responses of Sierra Club members were more divergent than those of either Earth First! members or sawmill workers, as listed in table 8.4. Our sample of Sierra Club members stands out in their reluctance to accept statements that might be labeled as speculative, pseudoscientific, or overly suspicious (1, 97, 56). They are most extreme in their rejection of a religious reason to preserve the environment (48), and with regard to energy sources, they are more accepting of nuclear power (127) and most likely to see a totally solar-powered society as wishful thinking (122).

While these divergent views of the Sierra Club chapter we surveyed may be surprising at first, they do make sense after further consideration. Of the groups in our survey, our sample of Sierra Club members are the best educated and oldest. These two characteristics may give our sample

Table 8.4
Survey statements for which Sierra Club members were the most divergent

Earth First!	Sawmill workers	Sierra Club	Survey statements
93	74	52	1 The weather has been more variable and unpredictable recently around here.
79	48	33	97 There may be a link between the changes in the weather and all the rockets they have fired into outer space.
94	59	41	56 The news the public gets on the environment is censored by the big companies and the government.
86	67	42	48 We don't have the right to play God by manipulating nature.
73	22	15	127 We shouldn't resort to nuclear power, even if the industry can make it completely safe, take care of wastes, and prevent it from going into bombs.
29	63	74	122 Believing that we can meet all our energy needs from solar power is just wishful thinking.

Note: The first three columns indicate the percentage agreeing with each statement.

of Sierra Club members a greater investment in society and its institutions. Consequently, in their survey responses they express a greater caution on scientific matters and a greater faith in government, industry, and technology. One lesson here is that even though a group may be selected for its position along an environmental continuum, that location will not explain all the differences among groups.

Conclusion

The patterns of agreement found in our survey results show that most Americans share a common set of environmental beliefs and values. Our analysis of the pattern of agreement leads to the conclusion that there is a single cultural consensus. Although some individuals appear on an antienvironmental end of the spectrum, these few individuals were scattered. Thus, in contrast to claims in earlier writing and earlier studies, we find neither a coherent and consistent antienvironmental position nor any other alternative position that leads to consistent answers on our survey.

A cultural consensus does not mean that each of our statements got either 100 percent or 0 percent from every respondent. In fact, looking through our entire set of survey questions in appendix C, 70 to 90 percent is a more common level of agreement. A cultural consensus means that there is only one set of culturally agreed upon answers. In other words, environmental beliefs and values are somewhat like geography (one set of facts, not universally known) or like etiquette rules (one set of norms, neither universally known nor always followed). Our finding means that the environment is not like the issues of abortion or gun control, both of which seem to have two alternative, coherent viewpoints.

The content of this agreement reflects components of widely shared cultural models and values, including those described earlier in the book. Statements that violate these environmental sentiments are rejected by most respondents. Marked disagreements were found only among our most extreme groups, and then only for a minority of statements that required choices between strongly held values, such as protecting the environment versus feeding one's family or allowing some humans to die in order to prevent species extinction.

The strong endorsement of environmental values by the diverse groups studied in our survey may well reflect a general willingness for the American public to make significant sacrifices for the sake of the environment. However, transforming this stated willingness into coordinated social action will not necessarily be easy. Policies must be crafted and leadership provided to overcome divergent individual and group self-interests.

9

Implications of Our Findings

Humanity has become the major force of environmental change on our planet, creating rapid and potentially destructive new global changes such as ozone depletion, rapid species extinctions, and anticipated global warming. The task of slowing these changes confronts us with an unprecedented challenge. All three global problems are long term and not directly visible, so to deal with them we must rely on sometimes-uncertain science and make value judgments about the future. Solutions are not impossible—a comprehensive remedy for the problem of ozone depletion is already underway. However, the problems of global warming and species extinctions will be more difficult to solve than was ozone depletion because they affect a broader range of activities and constituencies. Making the broad economic and social changes needed to solve these global problems will require not only the setting of policies and negotiation of international treaties by governments but also the understanding and active participation of the public. To investigate potential public support for environmental solutions, we undertook this study of the environmental beliefs and values of laypeople in the United States.

Our goal has been to understand the nature of lay American environmentalism. National surveys had already shown that environmental sentiments in the United States are strong and growing stronger, but we felt that those surveys did not really explain what American environmentalism is. We sought to identify the beliefs and values that underlie environmentalism, as well as how these beliefs and values are combined to form cultural models.

As anthropologists, we began this study not by defining environmentalism in order to devise survey questions but by asking laypeople to do so. Our semistructured interview format encouraged people to talk about the environment in their own terms. Our second phase of data collection, the survey, was built from statements offered by informants themselves in the semistructured interviews. We used this survey to examine the distribution of beliefs, values, and cultural models across American culture. These two methods complemented each other in that the semistructured interviews allowed us to understand lay perspectives in all their richness, and the survey measured their frequency and distribution across groups and allowed us to analyze patterns of agreement.

When we talked about the environment with people like teachers, coal miners, factory workers, custodians, and local environmentalists, we found that they often used language and concepts very different from those of environmental scientists and policy analysts. These differences were not just that laypeople have an incomplete or distorted version of the scientists' model but that they create their own cultural models that make sense to them.

Findings

We present here a sample of our major findings. Our informants see nature as a highly interdependent system in a balanced state, vulnerable to unpredictable "chain reactions" triggered by human disturbance. Global warming is understood using the prior cultural models of pollution, the ozone hole, and photosynthesis. The cultural models themselves are not wrong, but they are inappropriately applied to global warming, thus leading to incorrect inferences. Weather is considered susceptible to human influence, and a majority say they can already notice warmer winters.

We found that environmentalism has already become integrated with core American values such as parental responsibility, obligation to descendants, and traditional religious teachings. Surprisingly, biocentric values—valuing nature for its own sake—are also important for many.

People employ these values and cultural models in deciding which environmental policies they favor. For example, letting global warming proceed unchecked is seen as unacceptable because of the value placed on descendants. When most people use a cultural model inappropriately, majorities support policies that are unlikely to be effective. For example, banning spray cans and adding pollution controls were incorrectly seen as more effective than energy efficiency to prevent global warming.

When we examined individuals at length in case studies (chapter 7 and appendix D), we found that people fit together sets of cultural models and values in ways that make sense to them. Our case studies refuted notions of opposition to environmental laws as being due to a lack of environmental values or a lack of contact with nature. Rather, opponents of environmental laws typically share with other Americans the same cultural models and often the same environmental values, but those were overcome by competing models (e.g., believing that environmental concerns are politically exploited) or values (e.g., concern about human suffering, say, from coal workers becoming unemployed).

We quantitatively analyzed the patterns of agreement among individuals. Despite samples that overrepresent potentially antienvironmental individuals, we found a set of widely shared environmental beliefs and values. We found no corresponding cluster of people holding beliefs and values that were either in opposition to, or an alternative to, the environmental consensus. The few individuals who could be called antienvironmental at all did not agree with each other to nearly the extent that the proenvironmental individuals agree with each other. In short, we found a coherent, shared set of environmental beliefs and values, but the contrasting beliefs and values that might be an alternative are neither shared nor a coherent set.

In the first chapter we asked what is American environmentalism, and what causes high percentages of Americans to support environmental actions on national polls? We are now better prepared to answer these questions. In a sentence, lay environmentalism is built upon cultural models of how nature works and how humanity interacts with it, and is motivated by environmental values. Cultural models of the way nature works include the complex interdependencies among species and other systems, human reliance on the environment, and the way human activities affect nature. Environmental values include humanity's utilitarian need for nature, obligations to our descendants, the spiritual or religious value of nature, and for some, the rights of nature in and of itself. Finally, American environmentalism represents a consensus view, its major tenets are held by large majorities, and it is not opposed on its own terms by any alternative coherent belief system.

Our analysis of patterns of agreement also helps to make sense of the finding from national polls, noted in chapter 1, that 58 to 73 percent of Americans state "I am an environmentalist." Our graphs of agreement in chapter 8 show that two-thirds of the laypeople are in fact indistinguishable from members of a moderate environmental group like the Sierra Club.

Is the United States Unusual?

Our study examined lay environmentalism in the United States. A logical next question is, to what extent can our findings be generalized to other countries?

Two recent studies by Ragnar Löfstedt (1992; 1993) have applied our semistructured interviews, in Austria and in Sweden. Working with Kempton, Löfstedt developed semistructured interview protocols in German and Swedish that were very similar to our protocol. Löfstedt's work differed from ours in that his interviews did not pursue the general environmental questions as far, and his analysis focused almost exclusively on global warming. Like Americans, Austrians and Swedes both overextend their ozone hole model to global climate change (as do Australians, see Childs et al. 1988), and many also misapply the pollution model. Other aspects were different. For example, when asked open-ended questions about preventing global climate change, half the Austrians mentioned renewable energy or energy conservation, compared to less than 15 percent of Americans in our study.

Löfstedt's translated quotations from his interviews yielded startling similarities with those we obtained from our American informants. For example, "The greenhouse effect is coupled to ozone. It comes from the industry, the cars, and through [manufacturing] plants that have no filters" (Austrian locksmith). Or, in response to the question of why one should protect the environment: "It is our living space" (Austrian student). "If we destroy nature we destroy ourselves" (Austrian chiropractor). "It is very dangerous to change the system. There has to be some type of balance . . . within nature. Plants, animals" (Swedish business journalist). With minor wording changes, these could have been quotations from our American informants. Data from semistructured interviews do not lend themselves to a precise comparative analysis, but Löfstedt's quotations and analysis suggest that many of the American environmental models we discussed are held in other industrialized countries as well.

Outside Löfstedt's studies, our comparisons rely on data of less depth. For example, social scientists in Central and South America have anecdotally reported widespread concern about deforestation causing oxygen shortages.[1] This concern is presumably due to the overextension of a photosynthesis and respiration model, as in the United States. Concerning environmental opinion more generally, recent cross-national studies show that strong environmental concerns are expressed by citizens of most countries (Louis Harris 1989; Dunlap, Gallup, and Gallup 1992). Only a few countries show notably lower environmental concern, and a recent analysis shows that, contrary to stereotypes, environmental concern is not weaker in poor countries (Brechin and Kempton 1994).

General and Theoretical Questions

What general principles or questions follow from our study of American environmental thought? Claims of several earlier studies can be definitively rejected, and others can be at least tentatively confirmed. Given the diversity of cultural models and values we find underlying American environmental sentiments, we would reject the claims of some earlier studies that American environmentalism is based on a single component, such as health fears or enjoyment of the outdoors, or that it is generated primarily by the activities of environmental groups.

Other prior studies view public environmentalism as a paradigm or worldview. Does an American environmental paradigm exist? We found more direct evidence for environmentalism at a level of conceptual structure we call *cultural model*. Cultural models have a coherent logical structure and can be empirically identified with interviews or other methods. Scholars claiming to find an environmental paradigm have described a much broader set, which would include the entire collection of environmental models, beliefs, and values we documented in this book. Since our analysis of patterns of variation did find substantial sharing of models, beliefs and values, we would agree that it is reasonable to call American environmentalism a paradigm.

Our research also has implications for the study of science in society. We found that laypeople do not passively receive environmental news, but rather, they actively interpret what they hear via their preexisting cultural models. Our analysis of which cultural models were used for environmental issues showed why some lay interpretations are reasonable while others were inappropriate. (The issue is not whether the models do or do not correspond to scientific models but whether their inferences guide people to appropriate actions.) This process of science interpretation via cultural models means that one cannot understand laypeople's views of environmental problems, and presumably of other issues in science and technology, without first discovering the cultural models that underlie their views. The situation is somewhat different for specialists. (We use the term *specialists* to refer to nonscientists who know more than the average person about global warming because of their employment or political activities.) Specialists usually do not get environmental information directly from scientific literature. Rather, they get a filtered segment drawn selectively from the scientific literature, from sources such as colleagues, within-organizational documents, or interest group channels. Different specialists command very different sets of knowledge about the environmental problems we studied. As a result, when we interviewed specialists with opinions spanning all sides of issues, we often felt that their differential knowledge of the basic facts about environmental problems was much greater than their differences of opinion or of values.

One practical implication of our research addresses the question, do American cultural models of the environment work well? Our opening example of the Tukano showed that cultural models may guide people to protect critical environmental resources, even when these models do not correspond to Western science. This is also the case for some American cultural models, which can lead to reasonable judgments about the environment despite their lack of agreement with scientific models. A good example is the American model we found of biological chain reactions. Although this cultural model exaggerates species interdependencies, we argued (in chapter 3) that it leads to preferences for minimizing human impacts on species, preferences that are reasonable given the current rate of human-caused extinctions. By contrast, the ways in which current models are applied to global warming—for example, regarding it as an example of pollution or as being caused by the ozone hole—have led people to prefer ineffective consumer choices and policy responses over effective ones.

In these uses of cultural models, we found one apparent regularity in how well cultural models work. Environmental problems that are older in public exposure (pollution, ozone hole) are more widely known and are more likely to tap appropriate cultural models, thus guiding laypeople to reasonable conclusions. Newer environmental problems (global climate change, importance of habitat preservation versus saving individual species) are less widely known and are more likely to draw on inappropriate cultural models that can lead to erroneous conclusions.

To generalize, today's cultural models seem to be derived from the most widely publicized environmental problems of recent decades. For example, we presume that the environmental problem of insecticides traveling through food chains to multiple species has (in conjunction with other environmental controversies) stimulated laypeople to build the cultural models of pollution, species interdependency, and chain reactions. This apparent derivation of cultural models from publicized problems raises the question of how environmental models might change in reaction to today's global environmental problems like global warming. While we can only speculate on the future, we offer the following list of concepts and cultural models that might emerge: long-term sustainability, common assets of humanity, five-hundred-year time scales, intergenerational responsibility, and humanity's global interdependency.

In addition to these theoretical implications, our research also offers a methodological suggestion for future research. We began this study of environmentalism not by defining it ourselves but by asking a variety of laypeople questions like "Would you say that protecting the environment is important? Why?" Had laypeople's environmental thinking been captured in the literature, our labor-intensive semistructured interviewing would have been a waste of time and effort. It was not. Even for professionals who have been studying environmental problems for years, or perhaps especially for them, we found ample surprises.

We review a few of the unexpected findings. Counting only the non-environmentalists, 60 to 65 percent think climate change can be dealt with more effectively by pollution controls than by energy conservation; 80 to 90 percent believe that automobile and oil companies are in a conspiracy to suppress higher efficiency automobiles; 55 to 85 percent say they already notice the effects of global climate change on their local weather; and 70 to 80 percent say it is wrong to abuse the natural world because God created it (as do half of nonbelievers). These questions have not been asked on prior surveys, and we would not have asked them ourselves had we not conducted the semistructured interviews first.

Perhaps the most important question to emerge from our research is this: if American environmental values are so pervasive and strong, why is there not more environmental action? Why don't people act collectively to strengthen environmental laws, build infrastructure with reduced environmental impact, or institute incentives for behavior change? Or, why don't they act as individuals to reduce the aspects of personal consumption that are the most environmentally damaging? In our interviews, many do report making changes such as recycling and not buying spray cans. However, few are acting in the political realm, and few make major environmentally motivated changes such as moving to smaller housing or reducing the miles traveled by motor vehicle. The usual conclusion drawn from modest environmental actions is that modest actions indicate a modest commitment. But our research shows that this may not be true. The cultural models available to understand global warming, for example, lead to ineffective personal actions and support for ineffective policies, regardless of the level of personal commitment to environmental problems. Other studies have shown multiple additional barriers to action, from structural constraints (for example, that we have a road system but not a bikeway system) to fundamental incentive structures (for example, that companies earn more by producing and selling more, not by making people more satisfied with life (Stern 1992; Kempton 1993)). In short, for environmentally beneficial actions, environmental beliefs and values are necessary but often are not sufficient, given the multiple existing barriers to action.

Practical Advice

Based on our research results, we conclude with some practical advice, both for scientists studying global warming and for those—such as educators, science journalists, activists, or politicians—interested in communicating with the public.

Climatologists

Having conducted the research described in this book, we see a discrepancy between the questions about global warming that citizens want answered and the questions climate modelers are asking. Citizens want to decide about global warming based on the eventual long-term effects, as indicated in statements like Walt's saying the global warming effects would occur in our "kids' [lifetime] or grandkids' or great-grandkids'. It's going to be in somebody's lifetimes, that's for sure." When climate modelers analyze only short-term change, they are imposing values about the future that imply that only the short term is valued.

As we described in chapter 2, most studies of climate change have examined only CO_2 doubling, which would occur over the next fifty to a hundred years. Climatologists picked CO_2 doubling arbitrarily, as a common comparison point for discussion in the scientific literature. This prominence of doubling in climate analyses has in turn led to its prominence in policy discussions. For example, the Economic Report of the President cited in chapter 5 begins with estimated agricultural losses caused by climatic effects, discounts those damages to their present value, calculates that the damages from doubled CO_2 are not large enough to motivate action today, and concludes that nothing need be done about global warming. One problem with this analysis is that it assumes that climate change will stop at the level of CO_2 doubling. In other words, the benchmark has become, for policymakers, the only relevant future level of CO_2.

A more policy-relevant level would be the peak amount of CO_2 ultimately contributed by fossil fuels. Analyses by Sundquist (1990) and by Walker and Kasting (1992) suggest that the ultimate increase will be closer to sixfold than double. The consequences of a sixfold increase would probably be qualitatively worse than a doubling (Manabe and

Stouffer 1993). Given this, and our finding that citizens are concerned about a multigenerational future, we feel that the current practice of using the climatologists' benchmark of CO_2 doubling is misleading for both policy analysis and public debate. The choice of a fifty- to one-hundred-year time scale in climate models essentially precludes the value judgments our informants want to consider. Therefore, we would advocate more work by climatologists on realistic numbers to inform society about consequences of the actual CO_2 levels expected rather than more details on the partway point of doubling.

Teachers

Thanks to the earlier work of cognitive scientists like McCloskey, Caramazza, and Green (1980), some educational researchers are beginning to recognize the importance of preexisting cultural models in science teaching. Consequently, some new pedagogical strategies explicitly address those prior models. We add an additional point from our current research: there is more integration across models than previously recognized. For example, understanding of global warming is affected by cultural models of phenomena like photosynthesis, the oxygen cycle, and local pollution. To enable the next generation of students to understand global warming, teachers may need to draw out connections to these other models.

For instruction in ecology, our data suggest that it may be useful to distinguish between fragile interdependency and functional redundancy. In some cases, such as correcting the fear of oxygen depletion and qualifying the overemphasis on chain reactions, correcting models may reduce inadvertent environmental anxiety. We hope teachers would want to dispel overconcern, just as they would correct underconcern, to be true to their own educational ethics. But for any teachers who hesitate because they do not want to dampen student environmental concern, we remind them of the case study of Gerard in chapter 7, which suggests that well-intentioned exaggeration of environmental dangers can be highly counterproductive.

Our findings also offer potential links with the opportunity to teach multidisciplinary lessons. For example, discussion of the role of policies such as government regulation and environmental taxes could be integrated into units on the environment. We suggest this based on our finding that people support environmental taxes in general, but conceptual gaps—in subjects ranging from economics to chemistry—totally obscure how fossil fuel taxes would help prevent global warming. There are opportunities for creativity by educators and textbook writers in connecting these diverse subjects.

Communicators to the Public

Some communicators use public media not with the goal of educating but to persuade. We recognize that our findings could potentially be used for sophisticated deception. For example, environmentalists trying to build public concern about deforestation might raise fears of running out of oxygen. Carbon-emitting businesses could deflect criticism by reassuring the public about global warming with the claim that they have installed pollution control equipment and that spray cans are a greater concern anyway. Such exploitation of cultural models existed prior to our study, as seen in several recent environmental debates. Chapter 6 pointed out that opponents of the Clinton Btu tax exploited the cultural model of energy consumption as an unchangeable necessity. Another example occurred when the Bush administration promoted planting trees to combat global warming (implausible on quantitative grounds). President Bush appealed to the photosynthesis model to promote his program: "Trees . . . their beauty is breathtaking, and their bounty, breath-giving" (Gross 1990, 11).

Although we recognize that our findings have the potential for abuse, we hope that publishing them will have the opposite effect. By documenting the nature and importance of cultural models, this book should make it easier for others to reveal and refute such efforts to deceive.

Environmental Advocates

Our analysis of American values suggested that environmental advocates are missing an opportunity by basing their arguments primarily on utilitarian grounds (i.e., protect the rainforest because it has potential medicines). The only utilitarian value we found with real emotional force held that the earth should be preserved for our children and subsequent descendants. Two additional value bases for environmentalism are held by significant proportions of the population but are not now addressed by environmental activists: traditional religious teachings and an emerging biocentrism. Although these values are not as universal as utilitarianism, for those who hold them, they are stronger and more emotionally held. Environmental advocates who work in the political arena tell us that utility is still a more convincing argument for politicians—it may be hard for a representative to argue for the rights of other species when his or her constituents are threatened by unemployment. However, for those advocates who work with the public, appealing to a broader range of values offers the potential for a broader and perhaps more deeply felt reservoir of support.

Some advocates believe they still need to raise environmental concerns. We saw concern as already high and suggested that a greater problem is that the existing concern is often wasted on ineffective solutions, due to inappropriate use of cultural models. Can environmental groups address this? One simple, if partial, remedy would be to include specific policy recommendations along with messages about problems. A more complete remedy would require more comprehensive environmental education, which may be beyond the capabilities of environmental advocacy organizations and may be better provided by educators and the media. However, advocates can use their newsletters to steer their members toward more appropriate models, and, in the process of other publicity, they may be able to improve the cultural models used by the general public. As an example, a short video on global warming was recently produced by an environmental group. Intended for airing on television, it begins with pictures of factory smokestacks and dirty automobile exhaust, appealing to the incorrect cultural model of pollution as a cause of global warming. Better introductory visuals would be images of fossil fuels being extracted from the ground, to convey the idea that the quantity of fuel extracted and burned is a more direct cause of global warming than is visible pollution.

Anyone in the Political Sphere

Our research offered both good and bad news for those addressing environmental issues in the political arena. The good news is that in most conflicts over environmental issues there is more common ground than is usually acknowledged. One can assume, for example, that most people share the consensus statements in tables 8.1 and 8.2. The bad news is that people have serious misunderstandings about global environmental issues, which skew public support for policies for irrelevant reasons. When we have presented our results to government officials, environmental activists, policy analysts, and others involved with environmental opinion, some have concluded that trying to develop an informed public on such complex issues is futile. Instead of educating the public, this argument goes, we just need to pass the right laws and regulations. Without public understanding and support, however, we doubt that the political will exists to pass the necessary laws to deal with global environmental problems. The importance of public support can be seen in the history of the treaty on ozone depletion. As discussed in chapter 2, U.S. diplomatic support for the treaty was boosted by actions of consumers and voters. For problems like global warming and species extinctions, whose solutions will probably affect society more than those for ozone depletion, we argue that public support is essential for long-term success. Our research has shown that developing an informed public will be more complex and more difficult than is usually assumed. Our research has likewise produced some practical insights into how this effort might proceed.

Human societies have always had to adapt to changes in their environments. What makes our current situation unique is that the changes we confront today are global in scale, rapid, and caused by our own activities. Each of these characteristics makes them extremely complicated to solve. However, our findings provide cautious optimism in that the American public is already developing the values, and some of the conceptual tools, necessary for dealing with global environmental problems. Conceptually, public understanding of these problems is still inadequate, as seen in the substantial misconceptions about global warming—misconceptions that have already taken their toll in a lack of public support for environmentally beneficial policies. Nevertheless, current environmental values represent a major change in how our culture conceives of the relationship between humanity and nature. The speed of this change is stunning—it has surely occurred since the end of the nineteenth century and perhaps mostly over the past thirty years, since Rachel Carson published *Silent Spring*. Although we consider the incorporation of environmental values into our culture a significant positive step, the task of translating these values into effective action still lies ahead.

Appendix A
How We Gathered and Analyzed the Data

Purpose of this Appendix

Some readers will want to know more about our sampling, how we collected data, and how we analyzed it. This information is placed in appendixes to avoid interrupting the flow of the book for readers who do not need the details. This appendix describes interview procedures and sampling for the semistructured interviews, the sampling and analytical procedures for the later fixed-form survey, and the methodological background for the analysis of agreement and disagreement in chapter 8.

Semistructured Interviews

The first round of interviews uses semistructured interviewing, one of several ethnographic techniques that are valuable when the interviewer initially has little understanding of native concepts (see Agar 1980; Spradley 1979; Bernard 1994). Our semistructured interviews consisted of three parts. First, the interviewer asked questions about weather observations, environmental beliefs and values, and the informant's recall of information previously heard about global warming. Second, the interviewer gave a short briefing on global warming to provide background information. The briefing was designed to be similar in length and detail to an in-depth article in a weekly news magazine. Third, informants were asked for reactions to the briefing and to a set of policy proposals. (The full interview protocol can be found in appendix C.)

Forty-six semistructured interviews were conducted for this book, of which forty-three were transcribed and analyzed. The three not transcribed were interviews in which much of the conversation was off the topic and would have wasted expensive transcript and analysis time.

Interviews were conducted by one to three interviewers, most often by two interviewers. Two interviews were conducted by research assistants alone; the remainder were conducted with one or more of the authors of this book. The sample was divided into two main groups: *laypeople,* picked at random, and *specialists,* those whose work or interests relate to global warming. The lay interviews were conducted between July 1989 and July 1990. The specialists were interviewed between May 1990 and July 1991.

Twelve of the semistructured interviews were conducted in the near-urban community of Hamilton Township, a demographically diverse community bordering on Trenton, New Jersey. Four informants were approached in public places (a park or shopping center), and eight were approached in their homes. An initial two interviews were conducted with a married couple in Princeton, to test the question protocol.[1] In the home interviews in Hamilton, we picked houses by external appearance to span a diversity in neighborhoods and housing costs. Rather than prearranging interviews, we simply rang doorbells. After being warned that the interview might take 45 minutes, approximately two-thirds of those approached declined to be interviewed.

Although we felt this was a good mix of people, we did not want our study to be judged only on the first set of fourteen, drawn exclusively from a near-urban area in a state whose name is sometimes used as a synonym for environmental problems. We therefore conducted a second set of lay interviews in rural central Maine. The Maine interviewer (Hartley) grew up in the area and selected informants from acquaintances with a diversity of backgrounds. Two Maine informants, Frank and Ronald, were selected because the interviewer thought they would represent antienvironmental opinion. The Maine interviews also included some hunters and loggers.

Four specialist groups were chosen to represent people with a focused interest in environmental issues: environmental activists, coal miners, automobile engineers, and congressional staff dealing with environmental or energy matters. Five or six representatives were chosen from each specialist group. One person from coal and two from the automotive engineers were either lobbyists or very involved with compliance to government regulations (clean air, automobile efficiency). We found that these lobbyists, all environmental group members, and all congressional staff had a higher than average level of knowledge about environmental issues in general. Some of the congressional staff were experts on global warming in particular. The remaining coal miners, auto engineers, and two loggers could reasonably be assumed to perceive a personal cost to overly strict regulations on the environment, but were not experts on environmental issues. The environmentalists worked at the state or local level; none specialized in the issue of global warming. The congressional staff were selected to be evenly split between those supporting and those opposing current greenhouse-preventive initiatives.

Appendix B summarizes the demographics of the informants in the semistructured interviews. All informant names are pseudonyms. We offered informants a chance to make up their own pseudonyms; most chose to leave this to us. Demographic information is not listed for informants in some of the specialist groups, those for whom we felt we should be more careful about anonymity, but is used in our tabulations.

All but two interviews were tape recorded. (The tape recorder malfunctioned on one pilot interview, and one specialist refused taping.) All analysis of the semistructured interviews was drawn from our verbatim transcripts, occasionally supplemented by field notes. Most transcripts were made by a person present for the interview (book coauthor or assistant) and were checked against the tape by one of the authors. The transcripts total 530 single-spaced pages (376,000 words).

Quotations presented here use the following conventions. Quotes are word-for-word transcriptions from tapes, except that redundant "uh"s, repetitions, and false starts have been used more sparingly here than in a previous publication based on a subset of the data (Kempton 1991a), because we judged the detailed transcript form unnecessary and distracting to the reader. Details such as hesitation have been included only when significant to interpretation of the quotation. Nonstandard grammar and word choice are preserved, marked with [*sic*] when they might otherwise cause confusion. (Readers unfamiliar with verbatim transcripts should be warned that sentences from spoken conversation are rarely as complete and syntactically well formed as written text.) Ellipses (. . .) indicate material we have deleted. Brackets [ ] denote our postinterview clarification or paraphrase, as deduced from the context, from prior statements by the informant, from intonation, and so forth. Interviewer questions and statements appear in italics to distinguish them from informant responses. Brackets around our questions indicate a paraphrase of a long, rambling question.

The Fixed-form Survey

Designing the Survey

The fixed-form survey was developed based on the responses to the semistructured interviews. We first extracted what appeared to be the most important or controversial ideas from the 458 pages of the first 40 interview transcripts. Distilling the important points from the interviews yielded 165 pages of key ideas. We then continued to sort out these ideas until we had a core of 142 statements, selected to reflect the diversity of opinions expressed in the interviews. The statements express the informants' opinions, using close paraphrases of the informants' own words, and in many cases capture both the conclusions and part of the rationale for their judgments. In paraphrasing the informant, we tried to preserve the speaker's meaning even when the quote was pulled out of its context of the remainder of the interview.

In many cases, we deliberately preserved the speaker's language and advocacy slant of the original statement. Since this introduces a so-called bias into the questions, on important issues we included statements biasing it in each direction, for example:

109 Nature may be resilient, but it can only absorb so much damage.

Earth First!	Sierra Club	Public	Dry cleaners	Sawmill workers
94	93	97	93	85

Compared with a second statement, which similarly asserts that nature is resilient but draws a different conclusion from that fact:

29 The environment may have been abused, but it has tremendous recuperative powers. The radical measures being taken to protect the environment are not necessary and will cause too much economic harm.

Earth First!	Sierra Club	Public	Dry cleaners	Sawmill workers
0	7	23	17	33

Note that in addition to drawing a different inference, these two survey questions, like the quotations on which they are based, used different terminology (the "environment" versus "nature," calling existing environmental protection "radical") and slanted arguments to buttress their own beliefs. While neither is a "good, objective" survey research question, we find it more useful to know the range of agreement across the two questions, and to see that even when the position is strongly advocated, less than a quarter of the public accepts resiliency as an argument against environmental protection.

Chapter 2 gives an example illustrating the mapping from the semistructured interview transcript to the corresponding survey questions. In this process, we used some criteria that will be unfamiliar to survey researchers. For example, referring to the example on page 23, we worded question 105 "any human meddling," a stronger form of the assertion than Pervis's transcript statement. Our reason was that, since the semistructured interviews suggested that the belief in chain reactions was prevalent and strong, we used an especially strong statement in order to test a strong, deterministic form of this belief. Since majorities of all groups and three-quarters of the public agree with it even in this form, we were glad we used a strong form because it strengthens the result.

Another difference between our approach and most surveys is that our statements often combine two points. The above statement 29 combines an assertion about "recuperative powers" with a negative evaluation of current environmental measures. Like the above example, our survey's statements often try to build an argument, to see how many informants will subscribe to the whole argument. This is based partly on our observation that such statements were often embedded in lines of argument in the quotations from which they were drawn. A disadvantage is that in multipart statements we do not know whether informants are agreeing with every fact included, or more generally with the gist of the argument. Whether or not the reader agrees about which measurement technique is better, we note that our statements differ somewhat from the more isolated statements of most surveys.

In constructing the survey, statements from all the interview groups except automobile engineers were represented. (The engineers were interviewed too late in the study to include their interviews as sources of statements. However, in reviewing those transcripts, we believe we successfully included the most important propositions from this group as well.) Finally, the last seven statements were drawn from the "new environmental paradigm" questionnaire created by Dunlap and Van Liere (1978). These were added to enable comparison of our sample against earlier studies and because they seemed to capture background issues not specifically mentioned in the 142 statements derived from the semistructured interviews. (In retrospect, we feel our addition may have been imposing our view of environmentalism upon the instrument, even if to a far lesser degree than most surveys.) Although 149 seemed a rather large number of core ideas, we felt that trimming the number further would sacrifice critical points. Kempton and Boster distilled initial lists of statements independently; while combining them we were struck by how similar the selections were.

We then used these statements to construct a survey that asked respondents to judge on a six-point scale whether they strongly agreed, agreed, slightly agreed, slightly disagreed, disagreed, or strongly disagreed with the statement. Because we did not provide a middle neutral choice, we forced respondents to decide whether they agreed or disagreed with the statements, thereby allowing us to recode the responses to simple true false format. Very little information was lost in this recoding, as it correlated with the original six-point scale with a Pearson r of .997. Appendix C gives a summary of the original scale, for the interested reader. There were also three checklists regarding factual knowledge about global warming at the beginning and some demographic questions at the end.

The fixed-form survey had three parts. The first elicited the factual knowledge of the causes, consequences, and possible cures for global warming as well as opinions about the changes (if any) the respondent had observed in the weather. The second part elicited the degree of assent or dissent (on a 1 to 6 scale) to the 149 statements derived from the semistructured interviews. The third part elicited demographic information about the respondent.

Sampling for Each Survey Group

This section describes our sampling for each of the five groups in the survey. The Earth First! sample was taken by Bron Taylor of the University of Wisconsin, Oshkosh, during his attendance at two meetings of Earth First! in the summer of 1991, a national meeting in Vermont and a regional one in Wisconsin. Because he had to work hard to gain the confidence of the group, random sampling seemed impractical. He solicited respondents in two ways: indirectly by addressing a group, leaving surveys on a table, and asking for volunteers to take them, and directly giving the survey to individuals with whom he became acquainted enough to ask a favor. At each of the meetings there was a workshop on global warming, and he avoided sampling right after that workshop. Nevertheless, since this group constantly receives information about environmental issues, we consider it of minimal sampling bias—for this group—to sample some people who have heard a lecture on global warming within the past few days. Taylor's judgment is that the sampled group may underrepresent those who are anarchists or antiacademic, but that overall the sample is pretty representative of Earth First!.

The Sierra Club sample was taken by James Boster, who contacted a member of the board of the Orange County Chapter of the Sierra Club and asked for his cooperation with the research. Following approval at the board's monthly meeting, the chapter membership list was made available. Forty individuals were randomly selected from the list, and surveys were sent to each with self-addressed stamped envelopes. Twenty-eight returned completed surveys. This may be a reasonable sample of the Orange County chapter, but we would not claim it to be representative of the national membership.

The lay sample was taken by Karston Mueller, Jason Masterman, and Christina Rojas. Twenty-five people were interviewed in three California cities, using three methods of approaching the public. Seven people were asked to complete the survey during the 40+ minute BART (subway) ride between Walnut Creek and San Francisco. In Southern California, ten were interviewed during a door-to-door canvass of homes in Carlsbad (a middle-class neighborhood); eight more were interviewed on the beach near Huntington Beach.

Los Angeles area dry cleaners were selected because new pollution regulations will limit their use of perchloroethylene ("perc"), now the primary dry cleaning agent. This will require major financial investment in new equipment just to stay in business. The dry cleaner sample was taken by James Boster. Initially, firms selected from the listings in the Orange County yellow pages were telephoned and the managers were asked for their cooperation with the research project. Survey forms with self-addressed stamped envelopes were sent to the first forty managers who agreed to participate. Of these, sixteen had returned the survey after the third recontact. Subsequently, a second set of forty firms from the Los Angeles yellow pages were contacted and asked for their cooperation. Of these, fourteen returned the survey instrument after the third recontact.

The laid-off sawmill worker sample was taken by James Boster in cooperation with Randa Law of Lane Community College in Eugene, Oregon. This group was selected because it had been widely argued that loss of their jobs was due to the intervention of environmentalists to preserve old-growth stands of timber on behalf of the spotted owl. (Actually, an equally persuasive case can be made that the loss of sawmill worker jobs is due to the timber companies automating and shipping raw logs to Asia for milling rather than to U.S.-based firms (Meyer 1993).) Laid-off sawmill workers who enrolled in retraining classes at Lane Community College were asked to fill out our survey. Of the forty-eight surveys passed out to those who volunteered to participate, 27 were completed and returned.

Analysis of the Survey Data in Chapter 8

The goal of the analysis in chapter 8 is to determine the patterns of agreement among individuals and groups. The first step in the analysis is to standardize the respondents' responses on the six-point Likert scale so that all respondents' responses to the 149 statements have a mean of zero and a standard deviation of one. This reduces differences among the respondents in their different interpretation of the Likert scale both in response bias (agreeing with more of the items than other respondents) and in variance (using more extreme responses than other respondents). To accomplish this, we replaced blank responses with the mean of the respondent's own responses and then subtracted the mean and divided by the standard deviation of each subject. The standardization of respondents' responses recodes them to a scale reflecting the respondent's assent or disagreement with each statement, relative to the way they answered the other statements. A respondent-by-respondent correlation matrix is then created by correlating the standardized responses of each respondent with every other respondent.

The next step in the analysis is to test whether the data fit the cultural consensus model (Romney, Weller, and Batchelder 1986). The cultural consensus model formalizes the insight that agreement often reflects shared knowledge and allows the estimation of individual knowledge levels (cultural competence) from inter-respondent agreement. The central idea is that the agreement among respondents is a function of the extent to which each knows the culturally defined "truth." The application of their formal model requires that three conditions be met: that respondents share a common culture, that their answers are given independently, and that the competence of the respondents is constant over all questions. The model has been extended to ordinal and interval scale data (Romney, Batchelder, and Weller 1987; Weller 1987).

Minimum residual factor analysis of the respondent-by-respondent correlation matrix is used to check whether the conditions for application of the consensus model have been met. If the conditions are satisfied, there should be a single factor solution such that the first latent root (the largest eigenvalue) should be large in comparison to all other latent roots (at least two times) and there are no negative loadings on the first factor. The cultural competence of respondents can be estimated by the respondent's loading on the first factor, if the correlation matrix fits the consensus model.

We analyzed the responses of all informants ($N = 173$) and all items ($K = 149$) together, to see if a single belief system was present. In this case, the first factor was 5.6 times larger than the second factor, and the seven lowest loadings were $-.04$ to $-.26$, with all the rest between .07

and .85. In light of these results, we conclude that the model provides an adequate fit to the data. The informants below zero were only about 4 percent of the 173 informants surveyed, and all were from atypical groups that we overrepresented in order to find antienvironmental sentiment. Therefore, we conclude that there is a single belief system present, with approximately 50 percent of the beliefs and values shared. There is a single consensus, although there is still substantial variation among individuals. The factor scores are shown in table A.1.

This method was used to produce the figures 8.2 to 8.7. The horizontal and vertical dimensions are the first and second factor loadings of a minimum residual factor analysis of the correlations among respondents loadings on the fixed-form survey. According to the cultural consensus model (Romney, Weller, and Batchelder 1986), the first factor measures each respondent's shared agreement with the general consensus. The second factor reflects remaining differentiation among informants (Boster and Johnson 1989).

In chapter 8, we claim that survey respondent groups differ both in their environmentalism and their degree of agreement. This is illustrated in another way in table A.1. The column labeled "overall first factor"

Table A.1
Comparison of factor analyses on the overall sample and on each subgroup independently

		Overall 1st factor		Independent 1st factor		Eigenvalue Ratio	
Group	N	Mean	(sd)	Mean	(sd)	1st to 2nd	2nd to 3rd
Earth First!	31	.76	(.05)	.83	(.05)	37.40	1.21
Sierra Club	27	.67	(.14)	.68	(.14)	16.78	1.14
Laypeople	30	.58	(.21)	.60	(.19)	8.73	1.69
Dry cleaners	30	.57	(.20)	.60	(.18)	10.04	1.24
Sawmill workers	27	.38	(.29)	.49	(.19)	2.40	4.78
Semistructured	28	.50	(.31)	.52	(.28)	2.84	3.88
All groups	173	.58	(.24)	.62	(.21)	5.56	3.43

shows the mean and standard deviation of the degree to which each of the six subgroups share the overall belief system. The sawmill workers have the lowest average shared belief (.38) and the highest standard deviation (.29) of all the groups, while the Earth First! sample has the highest average shared belief (.76) and the smallest standard deviation (.05) of all the groups. This pattern remains essentially the same if independent consensus analyses are run for each subgroup. Although the average level of agreement increases in each group (e.g., the sawmill workers sample goes from .38 to .49, Earth First! goes from .76 to .83), the order remains the same across groups. Furthermore, the relative strength of fit to the consensus model also follows the same pattern: the ratio of the first to second eigenvalue is highest for the Earth First! sample (37:1), lowest in the sawmill workers sample (2:1), and intermediate in the Sierra Club, lay public, and dry cleaner samples (17:1, 9:1, and 10:1 respectively). The dry cleaners are essentially tied with the lay sample by any of these measures. This indicates that our results are robust, in that it means that each separate group fits the model (excepting the sawmill workers) and shares a consensus view of the environment.

To clarify the implication of not finding a clear and consistent alternative to the environmental consensus, we outline two possibilities the method could have produced, neither of which occurred. For this discussion, we use the term *consensus statements* to refer to the statements in our survey that large majorities answered identically. The first possibility is that there could have been an antienvironmental group of informants. Such a group would have agreed among themselves on the consensus statements but would have answered in the opposite direction from the majority. The second possibility is that there could have been an alternative view group who may have agreed among themselves on some of the consensus statements of everyone else but who had their own separate group of statements with which they agreed. That is, an alternative view group would have had their own separate set of consensus statements. This latter possibility was the case in the Tlaxcala-Pennsylvania example. It is not clear to us whether Milbrath, Dunlap, and others, as discussed in chapter 8, would consider their "dominant social paradigm" to be an antienvironmental or an alternative view. But neither view, in fact, emerges from our analysis.

To test whether there are systematic differences among the groups (which could exist even if there were an overall consensus), the Quadratic Assignment Procedure (Hubert and Schultz 1976) is used to compare the respondent-by-respondent correlation matrix with a structure matrix composed of blocks of ones (corresponding to within-group correlations) and zeros (corresponding to between-group correlations). Monte Carlo simulation (Hubert and Schultz 1976) is used to gauge the extent to which the correlation matrix and structure matrix are more similar than would be expected by chance. Matrices that are more similar than randomly permuted pairs in 990 or more of 1000 Monte Carlo trials would be taken as significantly similar (corresponding to a one-tailed probability of .01). This result would indicate the existence of subgroup consensuses as well: higher within-group correlations than between-group correlations. Although the lay sample was not significantly different from either the Sierra Club or the dry cleaner sample, for all other pairs of subgroups, members of the subgroups agree with others within the subgroup significantly more than they do with members of other subgroups ($p < .01$).

Appendix B
Informant Demographics

Purpose of this Appendix

This appendix provides background data on the informants from the semistructured interviews and the respondents to the survey. For both groups of informants, we summarize a few basic demographic variables to give a picture of the overall samples.

For the semistructured interviews, we also list each informant individually with background information. They appear alphabetically by pseudonym, rather than in the order interviewed. Some readers may find this background helpful in interpreting the quotations throughout the book.

For the fixed-form survey, we summarize the basic demographic information as a way of comparing the subsamples. The demographic information in the survey was very limited, both to shorten the survey form and because we intended to analyze across groups, not across standard demographic categories. As mentioned in chapter 1, public environmental concern has already been extensively analyzed by demographic category and found to have a strong relationship only to age.

Summary Background on Informants and Survey Respondents

The following two tables give summary statistics on the informants from both the semistructured interviews and in the survey. More background information was collected in the semistructured interviews, which were not as time constrained and in which we wanted to know as much as practical about the informants.

Semistructured Interviews

Table B.1 summarizes background information from the forty-three interviews that were transcribed and analyzed. The full list of informants follows as table B.3. Percentages are computed excluding missing values. "Voting Republican" includes those stating a party preference, plus independents who said they voted for Bush in the prior presidential election (1988).

Survey

Table B.2 summarizes some sociodemographic characteristics of the survey's five target samples. The other sample for the survey, a subset of those who already took the semistructured interview, was not analyzed as a separate group and is not included in the table. A question was added as to whether the respondent was a member of an organized

Table B.1
Summary characteristics of informants in the semistructured interviews

Group	N	Mean age	% male	Years school	Median income	% Voting Republican
Laypeople	20	47	65	12.8	$35,000	63
Specialists	21	41	71	16.8	$43,000	37
Pilot	2	35	50	18.5	$60,000	0
Total sample	43	43	67	15.0	$35,000	46

Table B.2
Summary characteristics of respondents to the survey

Group	N	Mean age	% male	Years school	% Organized religion
Earth First!	30	30	57	15	3
Sierra Club	27	49	70	16	41
Lay	29	34	52	14	34
Dry cleaners	30	40	73	13	50
Sawmill workers	26	41	85	13	33
Total sample	142	39	67	14	32

religion. We dropped an income question because we did not plan to analyze it.

Complete List of Informants in Semistructured Interviews

The following information will help interpret the list of informants in table B.3. The first column is the name, a pseudonym. The second column, "Group," is the sample group: L = lay people, La = lay, selected for anti-environmental views, E = environmental activist, CS = congressional staff, CW = coal workers, AE = automotive engineers, X = pretest.

For education, a degree is given (e.g., HS = high school graduate, BA = undergraduate Bachelor of Arts degree), or, for answers like "9th grade" or "3 years of college," the table gives the total years of school (e.g., 9 for 9th grade and 15 for three years of college). Income is reported for the household, whereas all other data are for the individual informant. The brief occupational description given in parentheses after each quotation in the book is abbreviated from the fuller occupational descriptions in this table.

In the table, "religion" labels the answer to "What is your religious background?" and "Are you active now?" "Political" refers to "How would you describe yourself politically?" (not necessarily a party) and their 1988 presidential vote.

The following codes are used when no data is available: (ref) = refused answer; (na) = not asked or unintelligible answer on audio tape; (anon) = information omitted here to further insure anonymity. "Anon" was used for informants who came from a small specialist group from which they might be personally identified. More variables are listed as "anon" for informants who were more concerned about being identified; for example, Mark and Gerard stated that what they told us they also said in public, and they saw no reason for us to make them anonymous.

Interview numbers 26 (Fred), 32 (Carl), and 41 (Randy) were not transcribed, do not contribute to analysis in the book, and are thus not entered in table B.3.

Table B.3
Informants in the semistructured interviews

Name	Group	Sex	Age	Educ	Household income	Occupation; race/ethnicity; religion; political; interview number
Abby	E	F	40	BA	$17K	small shopkeeper; white; inactive Catholic; independent, Jackson; 18
Alvin	CS	M	30s	anon	anon	legislative assistant; anon; anon; liberal Democrat, na; 25
Amanda	L	F	33	BA	$55K	housewife, was special ed teacher; white; active Protestant; Dukakis; 11
Andy	AE	M	46	MS	(ref)	government liaison, alternative fuels specialist; white; active Catholic; independent, Bush; 44
Bart	CW	M	34	law	$60K	lobbyist for coal-related labor; white; inactive Jewish; liberal Democrat, Dukakis; 40
Bert	L	M	44	HS	(ref)	proprietor of hunting and fishing camp, hunting guide; white; none; Republican, Bush; 31
Catherine	E	F	70	BA	$40K	retired school teacher; white; active Congregationalist; Republican, Bush; 34
Charles	CW	M	48	15	$35K	coal miner, operates machinery and does construction; white; inactive Methodist; Democrat, Dukakis; 38
Cindy	L	F	32	HS	$48K	housewife; white; active Protestant; Democrat, Bush; 12
Dan	AE	M	48	MS	$75K	technical specialist, fuel systems; white; Catholic; Democrat, Bush; 45
Doug	L	M	32	BS	$47K	pharmaceutical research; white; active Catholic; Independent, Bush “democrat for everything else”; 6

Eddie	X	M	36	PhD	$60K	university administrator; white; inactive Jewish; “liberal,” Dukakis; 2
Ellen	X	F	33	BA	$60K	freelance writer; white; inactive Catholic; “liberal,” Dukakis; 1
Emma	CW	F	39	MA	$31	coal mine loading-machine operator; white; agnostic; “far left or radical,” Dukakis; 36
Frank	La	M	58	HS	(ref)	building and construction contractor; white; none; independent, Bush; 30
George	CW	M	38	17	$33K	coal mine slate truck driver; white; inactive Presbyterian; conservative Democrat, Dukakis; 37
Gerard	CS	M	33	law	$43K	legislative counsel; white; active Catholic; conservative Republican, Bush; 24
Grant	AE	M	38	14	(na)	auto electronics engineer; white; inactive Methodist; independent, Bush; 42
Jack	E	M	29	BS	$32K	toxicologist for oil company; white; Presbyterian/agnostic; liberal Republican, Dukakis; 19
James	L	M	65	10	$22K	custodian, farmer; white; active Nazarene; Democrat, Dukakis; 27
Jane	L	F	81	HS	$14K	retired, clerical; white; active Protestant; independent, Bush; 13

Table B.3
(*continued*)

Name	Group	Sex	Age	Educ	Household income	Occupation; race/ethnicity; religion; political; interview number
Jenny	L	F	39	HS	$62K	high school teacher; white; active Catholic; “liberal,” Dukakis; 9
Joe	L	M	31	HS	$17K	factory welder; black, Liberian national; active Pentecostal; alien (nonvoter); 8
John	L	M	73	9	$16K	retired supervisor for wire manufacturer; Hungarian-born, naturalized; active Catholic; independent, Bush; 7
Kate	L	F	22	BA	(na)	student, starting to teach English; white; active Catholic; Democrat, did not vote; 28
Luke	CS	M	30s	anon	anon	professional staff; anon; anon; Republican, na; 23
Margaret	E	F	36	14	$30K	professional peace activist; white; raised atheist, sporadic Quaker; “not Democrat but sure ain’t no Republican,” Dukakis; 17
Marge	E	F	35	law	$50K+	environmental lawyer; white; not religious; liberal Democrat, Dukakis; 15
Marie	E	F	27	BS	$60K	county health inspector; white; Presbyterian; Democrat, Dukakis; 20
Mark	CS	M	40	(na)	$75K	legislative aide; white; agnostic; independent, na; 22
Mike	AE	M	50	BS	(na)	automotive systems engineering; white; inactive Catholic; “primarily Democratic,” Dukakis; 43
Nick	L	M	39	13	$80K	owner-operator, small manufacturing; white; inactive Methodist; independent, Dukakis; 16

Paige	L	M	39	HS	$53K	industry "production"; black; "non-affiliate"; na, probable Democrat, na; 4
Pervis	CW	M	38	15	$60K	coal mine wireman; white; active Methodist; Democrat, Dukakis; 39
Peter	L	M	46	HS	$30K	logging contractor; white; active Baptist; independent, Bush; 33
Ronald	La	M	59	BA	$50K	owner/operator of vacation cottages; white; Presbyterian; conservative independent, Bush; 35
Sam	CS	M	50s	anon	(ref)	senior professional staff; (anon); (anon); (anon); 21
Steve	AE	M	42	BS	$130K	coordinate auto models to meet CAFE; white; Catholic; conservative independent, Bush; 46
Susan	L	F	57	14	$26K	network administrator; white; active Catholic; na; 3
Tara	L	F	36	15	$40K	sales rep, sports equipment; white; inactive Baptist; "leaning Republican," Bush; 14
Tom	L	M	19	14	$20K	student; white; Baptist; independent, Bush; 29
Walt	L	M	73	HS	$13K	retired factory worker; white; active Catholic; na; 5
Wilbur	L	M	62	9	(ref)	retired fireman; white; inactive Methodist; independent, Bush; 10

Appendix C

Interview Protocol and Survey Questionnaire

Purpose of this Appendix

This appendix gives the full protocol used for the semistructured interviews and the complete list of statements presented in the fixed-form survey. As noted in the text, the precise wording and order of questions in the semistructured interview vary somewhat across informants, as is normal for this type of interview. For the briefing and the policy proposals, we did try to follow the protocol's written wording consistently.

A few questions were added to the semistructured interview during the course of the research. For example, we added more questions on environmental values after interview 14, and added question I.J on descendants near the end, after interview 39. The final version is included here, as it is the most complete and of greatest use for others who may want to reproduce our findings.

During the briefing session of the semistructured interview, we also used several charts, not reproduced here. They were a map showing expected temperature changes (which graphically showed that some places would become cooler, some warmer), a pie chart of greenhouse gas sources—duplicating the information on sources in table 2.2—and a map showing the geographical range of beech trees in North America with anticipated change from global warming. We also used a written list of predicted effects, which is incorporated into the protocol's text here.

For the survey we do not list the instrument itself. Rather, we list here the entire set of statements, along with a summary of informant responses to each statement.

Semistructured Interview Protocol

The complete interview protocol is reproduced below, as it was used in the semistructured interviews. This is our final revision of 10 August 1990.

0 Introduction and Permission

(lay) We're doing a study about people's opinions on weather and the environment. We are from ______University. Would you be willing to be interviewed? Most people say they find this interview interesting. It takes about 45 minutes, but you can stop anytime.

(specialist) This is a study of people's perceptions of the greenhouse effect. We are interviewing professionals dealing with this topic as well as ordinary citizens. The questions primarily concern your own personal opinions, but you are welcome to also mention the position of your office/organization if you wish. The interview starts with a few open-ended questions on the environment in general.

(both) Your answers will be held in confidence. It is faster if I tape, because I don't have to write everything as I go. Do you mind if I use the tape recorder?

I Current Model

I.A (lay) What factors would you say affect the weather? When you hear a weather report, what parts do you pay attention to?

I.B (lay) Do you think you would notice if one year was hotter or cooler than average? How would you notice that? Now we are finished with the questions on weather, and we are going to ask about the environment.

I.C (lay) When you think about the environment, what do you think of?

I.D Would you say that protecting the environment is important? Why? (If no, why not?)

I.E When people try to protect the environment, do you sometimes feel that other things do not get enough attention? (Prompt only if no response: "For example, some people say environmental protection shouldn't make people lose jobs and shouldn't raise the cost of living. How would you feel about those kinds of things?")

I.F Would you say that you have "environmental values"? (If yes) How would you describe those values?

I.G How would you describe the relationship between humans and nature? What should the relationship between humans and nature be? (If response: Do you have a spiritual basis, or a religious basis for that?)

I.H Would you say that people have a right to use some natural resources?

I.I Do people have responsibilities toward the environment? (If ambiguous: Is that for other people, or a responsibility to the environment itself?)

I.J (If not already mentioned) We've talked about protecting the environment for us today. Do you think it's also worth the extra effort and cost to protect long-term aspects of the environment, for our grandchildren or for other people who live in the future, even if it doesn't benefit us today?

I.K Have you heard about the greenhouse effect? (If no: Have you heard that there might be more hot weather in the future?) (If still no, skip to II.)

I.L What have you heard about that? (Probes, if needed: Have you heard anything else? What effects do you think that would have? Have you heard what causes it?)

I.M Do you remember where you got that information from?

I.N Do you believe this? (If yes or no: What made you decide that?) (If not sure: How would you decide whether or not you believe it?)

I.O If this happens, do you think it would be good, bad, or neutral? (If already obvious, put in form of verification: "So you think this would be a bad thing to happen?", etc.) Why? (Elicit what moral system is being applied.) (Possible probe: "Suppose we had the technology to change the climate for the better. Should we?")

I.P Do you think that something should be done about this? (If yes) What? (Prompt for all actors: government, corporations, individuals.)

I.Q (If I.P answered:) If those things are done, do you see any negative consequences? (Cover each answer in I.P) (If cost not mentioned: "For example, it could be costly to prevent global warming. Would you be concerned about that?")

(For lay, skip to II)

I.R Do you talk with other people at ______(your office/organization) about the greenhouse effect? (If yes: Is that something that is discussed a lot? What are other people saying?)

I.S Do you think the greenhouse effect will affect ______(your office/organization)?

I.T Do you think it will affect what you do personally in your job? (If yes, *explore fully.*)

I.U Do you know of people who you feel have a completely wrong approach to dealing with the greenhouse effect? (Who, for example?) (Why do they think that? What key points are they missing?)

II Briefing: Present New Information

Next I'm going to briefly describe what some scientists are saying about the greenhouse effect. I'm interested in any of your reactions as I describe this.

(specialist) We've tried to make this accurate, but this information relies on other scientists' publications, and we welcome your suggestions or corrections.

(Read)

Two different changes to the atmosphere have been in the news: ozone depletion and the greenhouse effect. The greenhouse effect makes it hotter, and ozone depletion allows more harmful rays to reach us from the sun. In this interview I'm going to cover only the greenhouse effect. (Lay first figure on table, or otherwise in view.)

(Show figure: Holding the heat) Naturally occurring gases in the earth's atmosphere trap heat near the surface of the earth, keeping it warm. This drawing illustrates the idea. Since the 1800s, human activities have been increasing some of these gases, especially carbon dioxide from burning fuels. Most scientists who have studied this change in the atmosphere think there is at least a fifty-fifty chance that it will raise average temperatures on the earth by about 3 to 9 degrees (F) during the next hundred years[1].

(Show temperature map.) This map was drawn by a NASA scientist to show his guess of what may happen in the United States. Redder areas indicate hotter temperatures, blue indicates colder temperatures. Temperatures may increase by different amounts in different places. At the same time, the soil may become dryer. In the midwest, hotter temperatures and dryer soil would make it hard to grow many of today's crops, leading to higher food prices. On the other hand, some parts of Canada might become better for farming.

Many scientists expect the greenhouse effect to make the ocean rise 1 to 6 feet in the next hundred years. A 1-foot rise in sea level erodes the coast 65 to 200 feet,[2] damaging coastal cities and destroying coastal wetlands.

II.A Do you have any questions about what I've said so far?
(Show Beech map.) Changes in temperature and moisture would affect many plants and animals. As one example, this map shows beech trees. Beech trees currently grow along the Eastern United States, from Louisiana up past Maine (point on map). The greenhouse effect may cause them to die out in all but the northernmost United States.[3] Beech is just one example. Other plants and animals in this area would move North, others would become extinct because they could not move or adapt quickly enough. Also, some plants, animals, insects, and parasitic diseases from the tropics might move into this area.

Climate change occurs naturally. For example, it is about 9 degrees (F) warmer now than during the last ice age, 18,000 years ago. But if the scientific studies are correct, the greenhouse effect would raise temperatures as much in only a hundred years, and if we continue to add these gases as we are today, it will just keep on getting warmer. Such a rapid change would be more difficult to adapt to, for human society as well as for animals and plants. The earth would become hotter than it has been in the past million years.[4] We don't know exactly what will happen.

(Show chart: Activities contributing.) Last, this chart shows what contributes most to the greenhouse effect.[5] The biggest part, over half, is burning coal, oil, and natural gas for energy. A smaller part is gases from what are called CFCs, used in refrigerators, air conditioners, foam insulation, and packaging. Others include cutting down forests, nitrogen fertilizers (N_2O), and methane (CH_4) from wet rice paddies and cattle. Most of these gases occur naturally, but human activities have increased the natural levels. What could be done about these gases? There is no need to stop using all these things completely. Scientists estimate that if we cut the amount of these gases we produce in half, we would slow down the process a lot. They estimate that cutting by three-quarters would stop it.[6]

(End read)

II.B Any reactions to that?

II.C Which things that I've described here were new?

II.D (lay) If you saw this information on the news for the first time, how would you decide whether it was something important to be concerned about?

II.E (Show list of possible effects) This page summarizes the main effects I mentioned. Which things would you be most concerned about? Why?

Possible Effects

1. Warmer weather, uncomfortably hot in summer.
2. Poorer farming, slightly higher food costs.
3. Sea level rises 1 to 6 feet. Coastal cities damaged, property lost underwater, and coastal wetlands destroyed.
4. Some plants and animals move northward.
5. Some plants and animals become extinct.
6. We don't know exactly what will happen. In just a hundred years it could be hotter than it has been in the past million years.

II.F Do you think anything should be done about this? (If yes) What?

II.G We've talked mostly about the United States. But today, the most deforestation is taking place in third-world countries. Over the next thirty-five years, third-world countries are expected to develop more industry, grow larger in population, and start producing more of these greenhouse-effect gases than the industrialized countries do today. What do you think could be done about deforestation and increasing emissions of greenhouse gases in the third world? (Follow-up anything indicating a moral system.)

III Policy Responses

Finally, I'm going to read five proposals that have been made as to what could be done about the greenhouse effect, and I'd like your reaction to each. We don't necessarily think all of these are good ideas, we're just trying to get a variety of opinions.

III.A One position is this: we cannot be 100 percent certain that the climate will change. Preventing climate change could be costly—why spend money if we may not even need to? Anyway, some effects, like warmer winters, may be beneficial. We could wait to see what actually happens and adapt to any changes when they occur. For example, if the Midwest becomes hotter and dryer, farmers could switch to crops that require less moisture, move north, or go into other businesses. Or, if sea level rises, populated areas could be diked, as has been done in Holland. What would you think about that?

III.B Another idea would be for government regulations requiring higher energy efficiency, to reduce the amount of fuel required by new products. The regulations might apply to new cars, houses, and appliances. Take cars, for example. Suppose new cars were required to get at least 55 MPG, thus cutting the amount of gas burned by half from today's new cars. However, let's say that this would mean that the next new car you bought would be a little smaller, and it would not be a high-performance car. What would you think about that?

III.C One argument is that people and companies use less when prices are higher. Therefore, one idea is to increase the price of fuels according to how much of these harmful gases they produce. Suppose there was a 100 percent tax on gasoline. How much do you usually pay when you get gas? (get $ figure) So a 100 percent tax would mean you would pay ______ instead of ______. What would you think about that?

Now suppose the same tax was on utility bills. What would you guess are your electric and natural gas bills (get total $). So a 100 percent tax would mean your bills would be ______ instead of ______. What would you think about that?

(If negative reaction) Suppose the 100 percent tax went straight into developing technologies that do not cause the greenhouse effect. Would this affect your reaction?

III.D Nuclear power does not produce the gases that cause the greenhouse effect. What would you think about a major expansion of nuclear power if that were a way to deal with this problem?

III.E Some people say that the best way to solve this type of environmental problem is with better technology; others say we need to change the way we live. What would be your opinion?

III.F Would you like to add any ideas of your own to solve this problem?

IV Background Information

Finally, I'd like to ask some background questions, so we can accurately sample a diversity of people. If there are any questions you'd rather not answer, just say so.

Age

How many children; age of oldest and youngest

Last year of school completed. (If > 12 yrs: What did you study?)

City of residence

Occupation

What do you do in your job?

What is your religious background? (Are you active (in ______) now?

What are your main sources of news?

Do you contribute to any environmental organizations?

How would you describe yourself politically?
(If not already given:) What is your party affiliation?
Who did you vote for, for president in 1988, when Bush ran against Dukakis?
How about in 1980, when Reagan ran against Carter?

What was your total household income last year from all sources? That would be the amount on your 1989 tax return before deductions.

Pseudonym

(specialists: get degree of anonymity desired)

V Informant Comments

What is your overall reaction to the things we have discussed today? (Probe: Anything else we haven't covered?)

(Record:)

M/F

race/ethnicity

how sampled

date, time, length of interview

where interview done

demeanor, dress, other observations

Survey Statements and Response Summary

Each of the statements in our fixed-form survey are given below, in the order they appeared on the survey. Most but not all of these appear throughout the text in sections covering related topics.

The two lines of five numbers below each statement indicate the answers given to the statement by each of the five groups. From left to right, the five columns represent answers by the Earth First!, Sierra Club, lay public, dry cleaner, and sawmill worker samples. Of the two lines, the top line is the percentage agreeing with the statement. For these percentages, as with most of the analysis in the book, we have collapsed "strongly agreed," "agreed," and "slightly agreed" together, yielding a single percentage of agreement, regardless of the strength of the agreement. The second line is the strength of the answers on a scale from 0 to 2, with 0 being either "slightly agree" or "slightly disagree," 1 being the middle strength, and 2 being "strongly" agree or disagree. We did not use the strength in the analysis in the text, as it doubles the numbers for each table and adds only a little information. Statements that approach 100 percent or 0 percent agreement also tend to have high strength values, and those near 50 percent agreement tend to have low strength values, so the agreement number alone contains most of the information in the data set. However, some statements and some subgroups vary from this pattern in interesting ways.

Responses to Fixed-form Survey

1 The weather has been more variable and unpredictable recently around here.

93	52	79	73	74
0.93	0.48	0.79	1.13	0.74

2 We are not going to be able to totally prevent global climate change, so adaptation has to be part of the solution.

68	78	67	73	74
1.10	0.63	0.67	0.73	0.81

3 If you don't appreciate the beauty of nature then you may not be as environmentally concerned.

83	59	73	70	44
1.03	0.93	0.67	0.90	0.89

4 If people only think of making a profit, they won't really see the beauty that nature has to offer.

100	78	86	87	69
1.55	0.96	1.21	1.23	0.85

5 Change of a few degrees in average temperature can make a huge difference in whether things can grow or not; global climate change would mean there would be enormous disruptions of agriculture.

100	85	100	93	70
1.77	1.15	1.13	1.10	0.70

6 We have a responsibility to use resources in the most efficient way possible without mindlessly depleting them.

94	100	100	97	100
1.68	1.78	1.63	1.53	1.15

7 People have a right to clean air and clean water.

94	93	97	100	100
1.84	1.67	1.60	1.87	1.44

8 How global climate change will affect me personally will determine how important a problem I consider it to be.

16	44	57	67	56
1.45	0.96	1.20	0.93	0.78

9 By the time the climate changes start occurring, it's going to be too late to do anything, the change will probably be irreversible.

50	33	47	43	26
0.77	0.89	0.73	0.73	0.74

10 The fact that there's so much unknown about global climate change is scary, it might be far worse than we realize.

94 96 87 90 74
1.52 0.85 1.07 0.77 0.67

11 Because global climate change is a world issue and not an individual one, its the government's responsibility to do something about it, not mine.

7 15 17 20 15
1.58 1.15 1.17 1.07 0.93

12 Our environmental priorities are mixed up, we ought to devote more attention to toxic chemicals and less to global climate change.

3 22 20 38 56
1.03 0.44 0.87 0.69 0.52

13 Global climate change has been blown out of proportion.

3 4 20 17 37
1.55 1.04 0.93 1.00 0.59

14 The young ought to be angry at previous generations over what they have done to the planet.

70 63 40 40 44
1.40 0.74 0.80 0.80 0.85

15 We're advancing so fast and are so out of control that we should just shut down and go back to the way it was in colonial times.

42 11 14 10 22
1.13 1.00 1.24 1.28 1.22

16 Justice is not just for human beings, we need to be as fair to plants and animals as we are towards people.

97 85 90 83 63
1.74 1.07 0.90 1.07 0.96

17 The rich are usually more wasteful than the poor.

90 70 50 53 63
1.39 0.74 0.80 0.80 0.96

18 The scientists have already done enough research on global climate change to start doing something.

94 78 87 70 78
1.55 0.89 0.67 0.60 0.81

19 It's the people who have a vested interest in exploiting the environment who say we have to wait for more studies of global climate change before we do something.

100 89 82 67 63
1.65 1.04 0.79 0.60 0.78

20 I would still buy the same amount of gasoline no matter how high the price went, it's a necessity.

16	37	53	60	44
1.65	0.63	0.57	0.80	0.78

21 We should be more concerned about the environment than the economy because if the environment is all right we can at least survive, even if the economic system is not in good shape.

97	85	73	67	59
1.61	0.63	0.77	0.60	0.70

22 In World War II, people gladly rationed for the war effort. People would willingly do it again to save the environment if the need were great enough.

77	67	57	83	93
1.13	0.74	0.87	0.80	0.70

23 People should pay the environmental costs of the things they buy. Products should be taxed depending on their effect on the environment.

97	85	70	77	48
1.58	0.70	0.87	0.67	0.89

24 We have to protect the environment for our children, and for our grandchildren, even if it means reducing our standard of living today.

100	100	97	87	74
2.00	1.15	1.13	0.87	0.74

25 We aren't justified in using resources to benefit only the current generation, if that creates problems for future generations.

100	100	93	90	82
1.84	1.37	1.10	1.07	0.63

26 The characteristics of cars are determined by the buyers, not the sellers.

23	56	60	52	54
1.13	0.78	0.53	0.90	1.04

27 We have a moral duty to leave the earth in as good or better shape than we found it.

100	100	100	97	96
1.66	1.37	1.17	1.28	0.74

28 We should invest in industry rather than spending money on the environment so that our economy will grow. Our children would then be more prosperous and better able to afford the cost of fixing any environmental problems we may have caused.

0	0	25	14	15
1.90	1.38	1.07	1.00	0.89

29 The environment may have been abused, but it has tremendous recuperative powers. The radical measures being taken to protect the environment are not necessary and will cause too much economic harm.

0 7 23 17 33
1.90 1.41 1.07 0.93 0.74

30 Why should the United States take drastic and expensive actions against global climate change when there are other countries that don't do anything. It has to be all countries or none.

0 15 20 25 52
1.59 1.11 1.07 0.96 0.89

31 Preventing global climate change is better than waiting to see what happens. It's more costly to fix problems than it is to prevent them in the first place.

93 100 97 97 85
1.76 1.41 1.03 1.28 1.04

32 It's unfortunate but acceptable if some people lose their jobs and have to change their line of work for the sake of the environment.

100 89 73 83 56
1.77 0.81 0.77 0.62 0.74

33 Energy conservation doesn't just make environmental sense, it makes economic sense too.

97 96 97 93 100
1.80 1.26 1.23 1.31 0.89

34 They could easily design a much more fuel efficient car, but the big automobile companies conspire with the big oil companies to keep that from happening.

83 82 80 90 89
1.47 0.81 1.10 1.17 1.41

35 Preventing species extinction should be our highest environmental priority. Once an animal or plant species becomes extinct, it is gone forever.

97 78 90 90 41
1.39 0.74 0.97 1.17 0.74

36 We're getting so far out ahead of other countries on environmental issues that we're crippling our economy and losing out to foreign competition.

7 15 23 37 59
1.67 0.85 0.67 0.53 0.74

37 We don't have to reduce our standard of living to solve global climate change or other environmental problems.

23 59 60 63 67
1.32 0.37 0.63 0.73 0.56

38 So far, global climate change seems to be only a theoretical possibility. We shouldn't make major changes in our economy and in our way of life for a theoretical risk.

0	7	24	13	48
1.77	0.78	0.72	0.80	0.48

39 A healthy environment is necessary for a healthy economy.

94	82	77	97	82
1.45	0.93	0.83	0.97	0.52

40 Communication and education are very important for dealing with global climate change. You've got to educate the young people who are going live in the future, both to help them adapt to the problems and to solve them.

100	96	100	100	100
1.61	1.52	1.37	1.43	0.93

41 Before the Reagan administration, they put money into making solar panels and wind machines, and other alternative sources of energy. Then all at once Reagan cut it all out and put all into the war machine.

97	78	63	54	67
1.33	0.59	0.77	0.61	1.07

42 Energy conservation will work better if we price energy correctly through higher fuel taxes to make efficient energy use in people's own interest.

100	76	73	50	30
1.35	0.76	0.70	0.67	1.04

43 We should reduce military spending and put the money into solving environmental problems.

100	82	69	57	63
1.84	0.59	1.03	0.77	0.93

44 The United States is not very energy-efficient compared to other industrialized countries.

97	42	61	40	52
1.39	0.50	0.71	0.40	0.78

45 I don't believe in preservation of species in the way some environmentalists do. In nature, evolution includes extinction.

24	48	37	40	74
1.45	0.59	0.90	0.77	0.85

46 The so-called experts on global climate change disagree so much with each other that it is hard to know what to believe. We really don't know whether there will be global climate change or not.

10	30	33	37	52
1.23	0.67	0.77	0.73	0.56

47 The way to solve environmental problems is to stop human activities that disturb the environment, and let nature heal itself, rather than actively intervening to "fix" the environment.

73	44	31	30	41
0.77	0.48	0.72	0.67	0.56

48 We don't have the right to play God by manipulating nature.

86	42	43	53	67
1.59	0.62	0.61	0.70	0.74

49 All species have a right to evolve without human interference. If extinction is going to happen, it should happen naturally, not through human actions.

100	82	87	77	59
1.81	0.93	1.03	0.97	0.78

50 Other species have as much right to be on this Earth as we do. Just because we are smarter than other animals doesn't make us better.

97	78	83	83	56
1.81	0.93	0.97	1.00	0.67

51 Our obligation to preserve nature isn't just a responsibility to other people but to the environment itself.

97	100	87	90	82
1.77	1.33	1.17	1.17	0.78

52 Religious leaders should try to do a better job of getting people to ask, "Is this the way God would have wanted us to treat the planet?"

100	85	79	87	63
1.37	0.77	0.86	0.87	0.78

53 The present relationship between humans and nature is one of domination rather than partnership. We look at most living organisms as extractable commodities.

100	82	90	87	81
1.81	0.81	1.03	0.83	0.58

54 Global climate change could cause humans to become extinct like the dinosaurs.

97	63	57	67	63
1.52	0.48	0.83	0.90	0.70

55 The reason politicians break their promises to the people to clean up our environment is the power of industry lobbies.

97	89	90	90	85
1.55	0.78	0.73	0.73	0.93

56 The news the public gets on the environment is censored by the big companies and the government.

94	41	77	77	59
1.16	0.48	0.57	0.63	0.78

57 Global climate change would disturb the whole chain of life.

100	85	93	90	82
1.84	1.07	1.00	1.07	0.81

58 Because God created the natural world, it is wrong to abuse it.

76	79	78	69	78
1.32	0.71	0.74	1.00	0.70

59 The third world has to deal with poverty and famine and can't even begin to think about environmental conditions.

29	44	63	60	50
1.19	0.48	0.67	0.80	0.77

60 Environmental degradation is the root cause of a lot of the poverty in the third world.

81	65	66	53	52
1.16	0.92	0.62	0.80	0.70

61 Maybe global climate change is a fulfillment of the biblical prophesy that the world will end in fire.

33	8	20	33	26
0.85	0.96	1.23	1.13	1.00

62 Poverty often leads to environmental destruction, whether in the third world or in poor areas of the United States.

83	89	70	60	52
1.07	0.93	0.73	0.67	0.70

63 My first duty is to feed my family. The environment and anything else has to come after that.

13	73	70	73	70
1.20	0.42	0.53	0.63	0.93

64 I would like to see private enterprise rather than government trying to find solutions to the problem of global climate change.

54	69	80	80	78
0.61	0.42	0.80	0.73	0.48

65 We spend billions of dollars trying to find the cure for cancer, and yet we're not doing enough to clean up the environment which causes the cancer.

100	78	80	97	78
1.58	0.81	0.93	0.90	0.74

66 Industry is the main group saying either that global climate change is not a problem or that the cost of dealing with it is too high.

100	89	90	79	74
1.61	0.85	0.70	0.62	0.74

67 Americans are too spoiled to change their lifestyle.

74	74	63	67	56
1.16	0.37	0.83	0.90	0.56

68 The public doesn't need to know all the details on policies to deal with global climate change, Congress just needs to pass the right laws.

29	22	27	20	19
1.19	0.78	0.97	0.93	0.96

69 The Creator intended that nature be used by humans, not worshipped by them.

0	30	35	52	59
1.81	0.74	0.88	0.72	0.81

70 Corporations and utilities have just about reached the end of what they are able to do for the environment.

16	7	17	17	8
1.81	1.26	1.13	1.03	1.19

71 If we could get people to change their lifestyle, we wouldn't need new technology to prevent global climate change.

61	42	38	43	41
1.00	0.58	0.55	0.47	0.70

72 We should be optimistic and assume that we are innovative enough to meet the challenge of global climate change.

45	70	73	77	63
1.03	0.78	0.60	0.57	0.59

73 An inherent flaw in our economic system is its inability to take a long time perspective on its actions.

97	85	83	87	73
1.77	0.89	0.73	0.67	0.62

74 The environment probably doesn't need as much protection as we imagine.

14	11	10	13	26
1.55	0.96	1.00	1.00	0.70

75 People's only responsibility to nature is to make it serve their own best interests.

0	0	24	17	11
1.90	1.44	1.03	1.10	1.00

76 Humans are ripping up nature, feeling that they can do a better job of managing the earth than the natural system can.

93 63 72 73 63
1.63 0.70 0.66 0.77 0.70

77 Humans should recognize they are part of nature and shouldn't try to control or manipulate it.

97 74 72 57 69
1.70 0.70 0.79 0.70 0.50

78 If any species has to become extinct as a result of human activities, it should be the human species.

80 22 21 33 7
1.53 0.81 1.03 0.93 1.48

79 The majority of people are completely cut off from nature. They spend their time indoors, and when they're outdoors, nature is just an inconvenience to them.

97 56 57 47 56
1.20 0.48 0.60 0.97 0.78

80 Plants and animals are there to serve humans. They don't have any rights in themselves.

0 7 23 10 31
1.87 1.07 1.13 1.03 0.81

81 There are too many environment regulations right now.

0 11 20 23 48
1.83 1.00 0.87 0.70 0.74

82 The problem I have with requiring fuel efficient cars is that I would resent the infringement of my personal liberty to be told I couldn't drive a particular kind of car.

0 19 27 30 41
1.80 1.04 1.17 0.83 0.67

83 Space exploration will probably be the final answer for this world that's being heavily polluted.

7 19 30 37 33
1.81 1.07 0.90 0.83 0.56

84 I would rather see a few humans suffer or even be killed than to see human environmental damage cause an entire species go extinct.

90 48 43 56 22
1.27 0.72 0.57 0.67 1.04

85 I can't do anything about global climate change. I have no power.

3 4 21 23 7
1.71 1.00 0.86 0.97 0.96

86 The industrialized countries have done the most to cause the problem of global climate change, so they should do the most to help solve it.

97	96	87	90	74
1.62	0.85	0.77	0.77	0.81

87 Most of the policies being advocated to respond to global climate change are things that we should do anyway, like conserving energy and developing solar and other renewable energy sources.

97	100	97	83	96
1.70	1.41	1.00	0.80	0.96

88 We should be skeptical about scientists' predictions of climate change. You can get five different estimates from five different scientists.

30	26	50	53	63
1.07	0.33	0.57	0.37	0.52

89 Unfortunately a lot of companies wouldn't do anything to protect the environment unless they were forced to by law.

97	100	97	93	78
1.70	1.19	1.27	0.97	1.00

90 Scientists are just speculating about global climate change. We shouldn't take action until they have proof.

0	0	21	20	22
1.87	0.96	0.93	0.87	0.70

91 The people that are concerned with global climate change are the same as the ones who are for nuclear disarmament. It is not a purely scientific concern, it is also very much intertwined with liberal politics.

37	41	40	33	52
0.90	0.67	0.80	0.47	0.74

92 A fuel tax would be an unfair way to reduce fuel consumption because some people are forced to use more fuel than others by their business or personal needs.

17	33	70	67	81
1.17	0.52	0.83	0.83	0.81

93 Global climate change may be too slow in coming to get people worried or interested.

72	70	83	63	67
0.72	0.63	0.53	0.57	0.52

94 If they cut all the forests down, we would soon run out of oxygen to breathe.

64	58	77	67	44
0.86	0.77	1.03	0.87	0.78

95 Global climate change makes life very exciting. It's like we're forging ahead into unknown, unexplored territory.

25 22 14 7 11
1.43 0.89 1.38 1.07 0.81

96 Some of the people who are the most passionate about global climate change are pretty naive about the facts. The ones who know what they're talking about tend to be cautious about making predictions.

25 70 62 80 70
1.04 0.41 0.55 0.30 0.48

97 There may be a link between the changes in the weather and all the rockets they have fired into outer space.

79 33 43 31 48
0.79 0.41 0.83 0.52 0.70

98 There are probably thousands of medicinal and other useful plants that are unknown to science that we might lose because of global climate change.

100 93 83 93 63
1.48 0.96 1.00 0.67 0.48

99 We should become vegetarians to reduce our environmental impact.

86 41 40 13 22
1.21 0.59 0.67 0.80 1.00

100 Plants and animals will successfully adapt to human-caused climate changes as they have to past natural climate changes.

17 48 33 33 59
1.38 0.44 0.63 0.53 0.63

101 Being out in nature can revitalize everything in you.

93 89 93 77 78
1.55 0.93 1.00 0.90 0.78

102 The government won't take care of environmental problems unless the people are concerned and are making a fuss about it.

97 96 97 90 78
1.48 1.22 1.30 1.03 0.93

103 You shouldn't force people to change their lifestyle for the sake of the environment.

0 0 27 13 41
1.74 1.07 0.90 0.70 0.59

104 If there is no economic, aesthetic, or other human use for a species—for example, some lichen in the desert—then there is no reason to worry much about it becoming extinct.

0 15 13 17 52
1.97 0.93 1.03 0.90 0.70

105 Nature has complex interdependencies. Any human meddling will cause a chain reaction with unanticipated effects.

97	89	77	76	63
1.77	0.81	0.77	0.52	0.48

106 Just because something is in the newspaper or on TV doesn't mean it is something to be concerned about. To be concerned, we have to see a problem with our own eyes.

40	19	30	20	33
1.07	1.04	0.93	0.90	0.70

107 Global climate change could destroy all human life.

90	67	67	73	56
1.39	0.70	0.93	0.97	0.78

108 If global climate change would cause things like a million people drowning in Bangladesh, we can't let that happen. Even if it doesn't affect us here, that would be totally immoral.

84	85	93	83	93
1.39	0.89	1.17	1.07	0.85

109 Nature may be resilient, but it can only absorb so much damage.

94	93	97	93	85
1.65	1.11	1.17	0.93	1.00

110 The most effective way to solve environmental problems is for individuals to stop buying environmentally damaging products.

55	82	73	87	70
0.77	0.52	0.83	0.63	0.67

111 Working to try to prevent environmental damage for the future is really part of being a good parent.

97	100	93	97	85
1.74	1.11	1.03	0.90	0.78

112 I'm not overly concerned about global climate change because it won't have any effects for over fifty years. I'm not going to be here that long.

0	11	0	0	15
1.93	1.19	1.38	1.33	0.93

113 There is no reason to be concerned about the effects of a 5-degree increase in average temperature because daily temperature varies by more than that from night to day.

0	7	13	13	26
1.93	1.22	1.23	0.90	0.67

114 There is nothing wrong with killing individual animals, as in hunting, as long as you don't kill so many that you threaten the population.

40	67	43	43	89
1.00	0.70	0.73	0.60	0.89

115 Energy conservation means doing without some things that give us comfort and enjoyment.

77	63	73	63	78
1.07	0.67	0.77	0.63	0.70

116 Reducing pollution is a more effective way to prevent global climate change than energy conservation.

21	36	66	59	67
0.96	0.52	0.55	0.45	0.48

117 As new technologies become available that are less environmentally damaging, companies will naturally want to adopt them and use them.

19	48	47	63	52
1.16	0.48	0.57	0.60	0.48

118 Before Columbus came to this continent, the Indians were completely in balance with their environment. They depended on it, respected it, and didn't alter it.

58	78	77	80	69
0.68	0.74	1.00	0.87	1.00

119 Capitalism may be the best system we know of today, but a fundamental problem with it is that it doesn't give any value to things you can't buy and sell, like the environment.

80	82	90	83	63
1.37	0.78	0.97	0.63	0.78

120 It's not inconsistent to be a hunter and a dedicated environmentalist as well.

74	67	73	60	82
1.29	0.70	0.80	0.63	0.96

121 Nature is inherently beautiful. When we see ugliness in the environment, it's caused by humans.

77	63	70	70	33
0.90	0.63	0.67	0.70	0.70

122 Believing that we can meet all our energy needs from solar power is just wishful thinking.

29	74	60	43	63
1.29	0.22	0.63	0.70	0.59

123 Environmentalists would not be so gung-ho if it was their jobs that were threatened.

23	39	48	37	78
1.10	0.50	0.69	0.47	0.81

124 Species of plants and animals have intrinsic aesthetic and spiritual value, even if they are not of any use to humans.

100 96 83 87 70
1.84 1.04 0.90 0.70 0.59

125 If we had the technology to change climate for the better, to improve the human condition, we should do so.

13 70 77 73 67
1.45 0.85 0.97 1.17 0.81

126 Global climate change would be bad even if it didn't cause humans any harm, because it is not a natural change.

94 74 87 83 67
1.42 0.48 0.83 0.87 0.70

127 We shouldn't resort to nuclear power, even if the industry can make it completely safe, take care of wastes, and prevent it from going into bombs.

73 15 40 50 22
1.30 0.65 1.13 0.70 1.07

128 Nuclear power doesn't pollute as much as other power sources. We should expand our use of it for the sake of the environment.

3 37 43 27 26
1.80 0.63 0.77 0.57 0.74

129 We should not force the auto companies to make cars with higher gas mileage. Instead we should discourage people's excessive use of their cars.

32 22 33 37 30
1.07 0.81 1.00 0.83 1.00

130 We shouldn't take action on global climate change based just on computer models and scientist's predictions. We should wait until we can measure an actual temperature change and know it's real.

0 11 27 20 30
1.77 1.04 1.03 0.80 0.81

131 Humans can't live without creating waste. We should use the old-fashioned methods—find some place out in the woods where it doesn't bother anybody, and just dump it.

0 0 13 10 0
1.84 1.44 1.53 1.23 1.41

132 We shouldn't be too worried about environmental damage. Technology is developing so fast that, in the future, people will be able to repair most of the environmental damage that has been done.

0 4 10 13 15
1.97 1.37 1.10 1.03 0.89

133 Scientists are always questioning and challenging previous theories. So they will never be 100% sure of global warming.

43	41	59	50	52
0.90	0.59	0.52	0.60	0.44

134 The average person can make a decision on whether they think we should do something about global climate change, even if they don't understand all the science.

94	96	93	73	78
1.26	0.67	0.97	0.67	0.67

135 You can already notice the effects of global climate change on the weather around here.

93	64	83	83	56
1.07	0.48	0.60	0.63	0.56

136 We really cannot prevent global climate change, because we'd have to shut down modern society.

13	15	23	17	26
1.53	0.74	0.70	0.77	0.67

137 The scientists who get into the newspapers and on TV are not in the mainstream with other scientists. They are on one fringe or the other.

23	11	40	20	48
0.90	0.81	0.57	0.67	0.48

138 There are certain basic necessities, such as heat, that people cannot give up no matter what the environmental costs are.

32	63	76	70	67
0.94	0.30	0.52	0.67	0.37

139 We can't expect poor people in this country to worry about environmental issues when their standard of living is so low.

32	30	40	30	33
1.06	0.52	0.73	0.67	0.85

140 Raising people's awareness of environmental problems is more important than getting the government involved in environmental regulations.

58	48	83	69	44
0.77	0.67	0.63	0.55	0.59

141 We should return to more traditional values and a less materialistic way of life to help the environment.

100	85	80	79	70
1.45	0.88	0.73	0.72	0.37

142 We should have started dealing with the problem of global climate change years ago.

100	93	87	93	81
1.63	0.67	1.17	0.93	0.65

143 Among the fundamental rights in the country is the use of ones' property without outside interference.

10	26	50	45	62
1.47	0.59	0.63	0.62	0.58

144 Government restrictions on the use of private property are necessary in order to insure that the land will not be permanently harmed.

90	93	72	79	67
1.20	0.70	0.97	0.75	0.74

145 The positive benefits of economic growth far outweigh any negative consequences.

0	4	37	17	19
1.70	0.96	0.73	0.59	0.56

146 The American people would be better off if the nation's economy stopped growing so fast.

87	56	43	28	41
1.30	0.44	0.64	0.69	0.59

147 Americans are going to have to drastically reduce their level of consumption over the next few years.

90	74	87	69	67
1.66	0.33	0.63	0.59	0.44

148 Regulation of business by government usually does more harm than good.

30	30	40	41	56
1.07	0.41	0.80	0.31	0.41

149 Government regulation and planning always leads to bureaucracy, inefficiency, and stagnation.

66	59	57	72	82
0.69	0.48	0.63	0.34	0.67

Appendix D

Case Studies of Citizens

The case studies presented in chapter 7 are individuals with particular interests in energy or environmental policy. In this appendix, we focus on three people whose jobs do not involve policy. This provides a more representative view of how individuals put together cultural models, beliefs, and values to make a coherent whole.

For these case studies, we selected people who span the range of environmental sentiment and whose interview statements contained more than the average amount of information. Thus, these individuals are representative in the content of their views but may be more thoughtful or articulate than average in expressing those views.

The three individuals are Tara, George, and James. Tara is a sports equipment sales representative who says she is environmentally concerned but has not acted on her beliefs. George works in the coal industry and is concerned about global warming strictly for its impact on humans. George is very focused on the trade-off between jobs and the environment; he is also fatalistic and suspicious of large companies. James is a small farmer and school custodian with deep environmental sensitivities.

Tara and George can be located in figure 8.2, showing the degree to which they share the consensus of environmental beliefs and values relative to other informants. George is far from the environmental center and Tara is within variation of the lay sample. Although we did not have survey responses from James, he would presumably be well within the more environmental side of the scale.

Tara (sales manager)

Tara is single, in her mid-thirties, with no children. She studied physical education in college for three years and is a sales representative for a golf clothing and equipment company. Politically, she describes herself as "technically independent but leaning toward Republican." Tara does not think of herself as an environmentalist but says she is "generally concerned" about the environment. She was interested in our questions and curious about global warming.

The following excerpts from our interview with Tara illustrate a common tendency to answer initially in utilitarian terms rather than revealing personal values. We quote at length from Tara's interview to illustrate the layering of justifications for environmental concern.

Would you say that you have environmental values? . . . Well, I don't know that I do. I'm not one, I mean not at this point have I joined any of these environmental groups, although I have thought about it. And I think I have a couple of applications sitting on the table right there, to send in my $25 or whatever. . . . I'm not taking an active part in it, but I'm generally concerned about it. *Well it isn't just whether you're active or not, but when you say you're "concerned" about it, what makes you concerned?* . . . The next thirty or forty years that I'm going to be on this Earth, I wouldn't want to think that I could die from air pollution . . . everybody should be concerned. . . . *Why . . .?* Well, for their health.

At this point in the interview, Tara's concern for the environment seemed to be based solely on protecting her own health. Later in the interview, she again referred to having thirty or forty years of life remaining, speculating that people with children would be more concerned than she is.

Let me follow up on that, let's just say, hypothetically: You don't have children; you know you're not going to have children. Would it still be of concern if we could say: "For thirty or forty years there's going to be very little effect, it's really something that's going to be happening" [informant completes sentence] "happening further down the line." No, 'cause that's being very selfish. . . . I mean all these years, the country . . . you take pride and like to talk history, where we started and how we began. And to think a hundred years from now, there's nobody here? . . . *Let me just ask why you say that. Why not be selfish?* It's just not the [pause] Well, I don't know, how can I answer that? Uh, that's just my opinion. . . . *I mean some people might say that goes against what my parents taught me, or . . . that's not what the Lord wants us to do. Do you have any sort of system you refer to when you think about what's right or wrong?* Oh, I see what you mean. Well, Christian. Uh, my faith, I guess if you want to say. . . . I think we should be concerned about the next guy, and I guess this was the way I was brought up in values and, "try not to be selfish," and that to me would be a very selfish way to be. . . . We're all really a product of what went on a hundred years ago, at least. And it's just sort of the nature of the way it's been and should continue.

This quote shows that when we pursued the topic (more doggedly with Tara than with most other informants), her initial value-free statements, "I don't know that I do [have environmental values]" and "I wouldn't want to . . . die from air pollution" gave way to a mix of value systems: American history, national pride, family teaching not to be selfish, and Christianity.[1] These values would not have been brought out had we not followed our initial question with probe questions, as part of the semi-structured interview format.

Tara clearly has feelings about what is right and wrong, although she has trouble justifying those feelings because she has not thought them through. She careens through multiple justifications rather than connecting to any coherent system. Also, her responses such as "just my opinion" seem to reflect a belief in cultural relativism—that there is no objective basis for morality.

Tara mentioned that just after she had first seen a television program on global warming and become concerned about it, she met someone through her work who was very worried about the issue. He gave her several membership forms for environmental groups. If she had more money and more time, she thinks she would join one.

In response to our question *"What have you heard about the greenhouse effect?"*, Tara says "That's where, ah . . . that has to do with the ozone doesn't it?" Tara uses both ozone depletion and photosynthesis to understand global warming. She says of global warming, "cutting down all of our woods takes away . . . the oxygen or something." She believes that more of a greenhouse effect would cause winters to get colder and summers to get hotter. She says that her information comes from local papers and, occasionally, the *Wall Street Journal,* but the source that made the biggest impression on her was the TV special she saw, which she thinks was on PBS.

As an "outdoors person," Tara would find the effects of global warming personally "very upsetting" in that it could affect her outdoor activities. After the briefing, she is uncomfortable with the idea that global warming could bring changes in many areas of life.

> It's just a scary thing. You're not going to have what you are accustomed to. . . . We get used to living a certain way . . . it'd be a major adjustment. And I don't know if a lot of people could take it either.

She cites the introduction of new parasites and diseases, as well as the unavailability or higher prices of common food items, as examples of changes that would particularly disturb her.

She believes that all the effects of global warming are worthy of concern and thinks the government should do something about the problem, but she does not feel qualified to offer any concrete suggestions, saying, "That's why we have these people in office." She feels that automobile mileage regulations would not affect her much, but she is wary of government regulations in general.

> I guess you're getting into what this democracy too is all about. They're gonna tell you that you can't drive a certain car, and I don't know if I go along with that. . . . I just don't know how many people would like the government telling them what kind of car to drive. In the Soviet Union that's another story.

She said she would not like fuel taxes because they would make a serious dent in her budget, but would accept fuel taxes "if it could help."

Tara believes that a combination of new technology and changes in lifestyle will be required to deal with global warming, but her primary concern is that people become educated about the problem.

> You know, I think the people have to learn. People have to be more aware like [I have become.] A lot of people like myself have to learn just a tiny bit.

She also wants to see more government interest in the issue.

> I think the government should be as concerned as the environmental groups. . . . I don't know how they lobby, you know, or what they're doing in terms of Washington, but I think they need to get closer in touch and strike somebody's bell in there to get the concern going.

Tara says that she is concerned about the environment and is interested in learning more about environmental issues. However this has not, to date, translated into even minimal action. This comes across in her description of being concerned about new reports on global warming. She recalled that her reaction was "Oh my God, this is really going on.

What can I do to help?" She had an opportunity to do something soon afterwards.

> I was calling on an account, and a customer just happened to be standing in the store when I was [there]. We were talking about this program that I was watching [about global warming]. And the next time I went [there] he had a couple of applications [to environmental groups]. *So he gave you those?* Yeah. He was really into this, so that was kinda fun. But I've, to tell you the truth, I've kinda forgotten about it, with all that's been going on in my life lately. *Sure.* I have too many more serious things really. At present. But it's something, I suppose, if I had a little more time and I had a little money, I would probably, probably get somewhat involved.

Tara does seem to have a feeling of responsibility to future generations, despite not having children of her own. She values the outdoors, at least for recreational use, and is concerned about global environmental changes. Further, she believes she should be devoting more energy to environmental issues than she currently is. These factors may explain why Tara is accepting of policies to deal with global warming—she even accepts the idea of a 100 percent energy tax. She has not made a personal commitment, however. Despite having an opportunity to take a small action (joining an environmental group), she has yet to take the step from being interested and generally concerned to personal action.

George (coal truck driver)

George is a slate truck driver working for a West Virginia coal mine, where the coal industry has been adversely affected by environmental legislation. He is a member of the board of directors for the Southern Appalachian Labor Board School and is working on a master's degree in labor and policy studies. He describes himself as a conservative Democrat. He is thirty-eight years old, married, with one young child. Unlike most of our informants, George was skeptical about the value of the work we were doing, as well as his own potential to contribute to that work. He questioned us about our work extensively, both at the beginning and end of the interview, and clearly believed that we represented the environmentalists' position on the global warming issue. He was interviewed by Boster and Clark in an ice cream shop.

Answering our first question regarding the importance of protecting the environment, George presented an outline of his views, based heavily on economic factors.

> [Protecting the environment] is important, but there's other aspects too. I mean, it's not just one, simple answer. . . . You have to have air to breathe, you have to have land to live off of, you gotta draw resources from the land. But also . . . people have to work to make livings. And if they don't work to make livings—it's like a catch 22.

George continues to bring an economic focus to the issues discussed throughout the interview, one of the most consistent to do so among our informants. He believes environmental protection is important because people need the environment to survive but sees environmental issues frequently conflicting with people's economic survival. He tends to see environmental and economic concerns starkly at odds.

> Give up your job or save the world. I mean, which one are you going to do? If you save the world, then you'll starve to death. So what's the use of living? I'm going to die anyway.

George indicates that although his associates are aware of environmental concerns, these issues remain nebulous, hypothetical threats. What people are really concerned with, he says, is "the economy, jobs, livelihoods, health care, things like that." Global warming cannot at this point hope to compete with economic concerns for people's attention because it is not yet a reality for them. "Nobody is really concerned about an issue until they confront [the issue] or it makes them [confront it]." George discusses at length the tremendous loss of jobs created by increased mechanization of the coal industry over the past ten years, emphasizing that this job loss is the focus of West Virginians' attentions. When asked if people have a sense of the potential problems the greenhouse effect poses for the coal industry, he draws a comparison with the acid rain issue.

> Nobody thought about acid rain as far as working on the local levels, I would say. The international people in the organizations, they understand it, and they understand the impact that's going to [have] if they don't have something to protect them. But the people . . . on the mine force or at the work place, until it actually hits them—then they [react.]

George recalls the words of a former professor: "Don't worry about things you can't change, worry about things you can," adding, "If you sit around and dwell on things that could happen, or would happen, or might happen, then you'd never have time to do anything else." He says his own environmental philosophy is similar to those of his cohorts, admitting that he takes the environment for granted. "Everybody else does." He takes a rather dismal view of the environment's condition and an equally grim view of his or anyone else's ability to change it. He repeatedly returns to the idea that people could, and perhaps should, react to problems before they become too acute to solve, but they cannot lift themselves out of their ruts to take action.

> [The environment] is just like dying. It's just like a mother or whatever. You never go visit your mother or your father because you're busy doing this, you're busy doing that, you're working this and that. And then when they die, you say, "Oh, I wished I would have done this, I wished I'd done that."

George had difficulty answering our question about what the relationship between humans and nature should be. In keeping with his general theme, he remarks that he has never been faced with that question before, so he has never considered it. After some thought, he said that people should put back what they take away from the environment and have more "self-conscience" about their actions. He quickly adds that this is not, in fact, the case. "I don't argue religion," says George, ". . . because it's too easy to get in too many arguments." He does believe people would have a moral responsibility to the environment in a perfect society, but points out that ours is not perfect.

> You're asking me to say morally and ethically if somebody, if everybody should say, "Yeah, I should . . . take better care of this and that." But they don't. They're not going to do it, and there ain't no way that you can make them do it.

George concludes his discussion of the relationship between humans and nature with a final dose of fatalism.

> A lot of people say that when it's your time to die, it's your time to go, and [it] ain't going to make no difference what it is or where you're at. I mean, that's it. And whether it's by the environment or whether it's by a car crash, whether it's by airplane or lightning striking you, when it's your time to go, it's your time to go.

George does believe that those who blatantly and knowingly abuse the environment should have to suffer or make retribution for their abusive and negligent behavior. He cites a New York State oil company that has for years knowingly allowed a huge underground oil leak to continue. The potential damage to human life from such corporate behavior incenses George.

> You talk about the earth as if the earth was alive . . . but *people*—it's a blatant disregard to the people that live on the earth.

George exhibits far more emotion over abuse of the environment by big business than by individual people, saying:

> It's bad enough that people abuse, but it's even worse when you have corporations or these energy conglomerates, or whatever it is, that abuse the environment. . . . That to me is the big issue. . . . Those people should have to be held responsible for it. . . . They have a responsibility to other living human beings.

George's information on global warming comes from congressional reports and news magazines. He shows the mix of models for the greenhouse effect and the ozone hole noted in chapter 4. The ozone layer is deteriorating, caused by burning fossil fuels, and when the ozone is depleted, he says, the earth gets heated from the inside because the heat has no way to release itself. The UV rays, he has read, will be so hot that they could, potentially, "fry people." He is not sure he believes everything he has read because he knows people are paid to write articles from certain perspectives and he has not personally observed any significant indication that the climate is heating up.

If global warming is, indeed, a genuine problem, George believes it is the government's responsibility to deal with it. He is vague about what the appropriate measures might be: "whatever's appropriate, whatever's deemed . . ." He remarks that he is not a scientist and does not know what steps should be taken, but he was unfavorably impressed by news he had heard shortly before the interview, reporting then-President Bush's lack of support for European resolutions to combat global warming.

George would support MPG standards as long as large families would be able to have cars big enough for their needs. He would not favor taxes on either gasoline or utilities, saying that such taxes would make it difficult for some people to cover their expenses. If such measures were taken, however, he would "go with the flow like most people do." He believes a combination of technology and altered lifestyles would be needed to solve the global warming problem but points out problems with both approaches. He has little faith in people's ability to make dramatic changes in their lifestyles on a long-term basis.

> It's like going to a doctor, and the doctor tells you need to take vitamins. You might take them for a week, and then after that, you don't take them anymore. So that's the way people would be if you told them to change their lives. They might change for a while, but after a while they'd go back to the same thing.

According to George, new technology would be easier to implement than lifestyle changes, but he is cautious about the uncertainties involved. He believes potential risks from new technologies are often hidden for political reasons.

George ultimately believes that at this point the "wait and see" policy toward global warming is best.

> It puts me in the mind of . . . watching a cow that was ready to have a calf. You don't do anything until the time comes. So you just keep watching. I think that would be sensible. . . . I'm a conservative in a lot of my thinking, and . . . I don't think that running off on the deep end is the answer.

Due to the conflicting scientific evidence, he is unconvinced that there is a problem and believes that the whole issue is so intertwined with politics that it will be difficult to achieve agreement on it in the near future.

> Who's to say who's right or wrong? I mean, that's the big problem. Everybody's supposed to follow what your president says. . . . There's a Republican president in there, Republicans think he's right, Democrats think he's wrong. It's just vice versa if it's the other way. So no matter what they say or what they do, or what their policies are, [to] a lot of people . . . it's right. And to other people it's strictly wrong. We'll find out sometime.

George does not see global warming as a pressing problem at the moment. Should the evidence become more convincing and point to a definite problem in the future, he would view the problem strictly in terms of the economic difficulties that changes in industry and lifestyle would create. Even if global warming becomes an issue of concern for all scientists and politicians, he does not predict that the average working citizen will be willing to make sacrifices to combat the problem because of what he sees as people's inherent inability to deal with a problem until they are face to face with its consequences. Until people are actually affected by global warming, he concludes, they will "just go home and watch TV."

James (farmer, custodian)

James has spent all of his sixty-five years in rural Maine, moving in his early twenties to the farm where he and his wife have raised seven children. Although clearly intelligent, he never went beyond the tenth grade, a fact about which he is obviously self-conscious as he expressed his opinions in an almost self-deprecating manner. To support his family, James has worked for many years as a high school custodian, a job he enjoys, but as he sat with Hartley for the interview in his kitchen, he was every inch a farmer. James is earnest and thoughtful in his responses to our questions and eager to learn more about global warming through our briefing. He describes himself as a Democrat.

James expresses deep environmental concern, saying that the way in which his farm ties in with so many aspects of the environment has made him very aware over the years of the environment's importance and fragility. It has also shown him how interconnected everything in nature is. In addition, he says that as he has grown older, he has increasingly realized the importance of passing on a healthy planet to future generations. James sees two main connections between humans and the environment: he believes that contact with nature keeps people physically and mentally healthy, and he sees a spiritual link with nature in that God created the natural world for humans to enjoy, and humans in turn have a responsibility to take care of what God has provided. People have a right to use these gifts but not to overuse them to the point where they are irreparably damaged or no longer exist.

> God has put us on this earth, and he has produced so many marvelous things for us that we tend to neglect and not take care of.

James believes a balance must be maintained between human needs and the needs of nature to prevent the ultimate destruction of both. He responds to the suggestion that some people could lose their jobs because of environmental protection, saying:

That's true, but, you know, I find that if you're going to take care of something, which is important to our livelihood, you've gotta swap one good for the other. You pay the price. There's no question about it. And I feel that is one thing that's caused a lot of problems in these last few years of people not wanting to cause any problems, so they've let things ride, and to me that's not the way it should be, because in the meantime it's getting worse and worse.

James gets his environmental information from newspapers, magazines, and television. He maintains that global warming would make winters become warmer and summers become cooler. He has read about the cooler summer prediction in farming magazines and is concerned about the impact this would have on crop growth. He finds support for this theory in the fact that he has noticed that the past few summers have been unusually cool. Warmer winters are also linked to this theory, based on personal observation of weather over the years.

James is amenable to all of our policy proposals, acknowledging that higher taxes and higher utility bills would be a hardship, but he is concerned enough to make the sacrifice. He expresses great doubt, however, about whether the government will take any action on global warming in the foreseeable future. He believes that preventive action should be taken now rather than waiting to adapt to change but does not see either the money or people with the inclination to do so.

It would be better [to spend money now on preventative measures] . . . but I think probably there aren't enough [people who] could care less what's going on. . . . But once we were in a sad situation and things were not doing the way we should, they'd be the first one to complain.

James has little faith in the political process. He has been disappointed too many times by politicians who make promises at election time, only to forget them once elected.

I used to [have faith in government], but I don't any more because they will put themselves before us.

He sees industry putting a strong check on governmental effectiveness in dealing with environmental issues.

[Industry is] so powerful. They lobby so hard. I think that that's probably one reason we haven't gone as far as we should.

James believes in the potential of the American people, government, and industry to deal with the problem of global warming, but expresses exasperation at what he sees as misdirection of energy and lack of concern with global warming and environmental issues, generally.

You know, it's strange with the smart people we have in our country—of all the things, they put man on the moon, and they do all these things for the atmosphere. But, hey, we got a problem right here. And to me if we continue to keep going up there we're gonna foul that up too . . . let's take care of what we got here and use it wisely, and see that it's taken care of the way it should be. . . . And look at the money being spent . . . I sound a little as though I'm against space. I'm not. But for the millions and billions of dollars being spent on it, we could use some of that right for keeping our own earth cleaned up down here. Maybe I'm wrong.

James believes that every person can have an effect on the world. "Each has to do his own part in order to keep the environment the way we should have it." He once believed that politicians worked for the people and that the political process worked in harmony with this idea, but in recent years he has soured on government. He is no longer sure that the people can actually have an impact on government. Even if there are good representatives who listen to the people other than at election time, he feels their power is limited by the special interest groups that he sees the government now walking hand in hand with.

You can write to your senators and congressmen. If you've got a good one and a guy that, well, for example, George Mitchell. I really and truly feel that he's been trying . . . on the acid rain bit . . . to do his best. But one cannot do it alone. Now whether or not if all of us got together and said, "Mr. Mitchell, we are behind you and we want you to do this and that," I don't know whether he could have the power to do it or not . . . whether he would be able to go far enough. Politics are funny to me, and geez, I've changed my attitude in the last few [years] about it, I'll tell ya.

In sum, James is deeply concerned about the well-being of the environment. He sees environmental health as essential to human health physically, emotionally, and spiritually and he wants to preserve a healthy earth for future inhabitants. He also respects nature because it was created for us by God, and we therefore have a responsibility to care for that creation. James recognizes that environmental protection requires sacrifice, and says he is willing to make sacrifices himself. He is disturbed by the fact that not everyone shares this belief and is particularly troubled over what he sees as the growing separation between the American people and their government. At the same time he sees a strengthening of ties between government and industry. James wants to believe in his ability to contribute to positive change in environmental matters but is worried that his own actions may be eclipsed by a government that does not care for its people any more.

Notes

Chapter 1

1. We sometimes use phrases like "American environmentalism" to avoid wordy locutions like "environmentalism as subscribed to by residents of the United States." We are aware that outside the United States, "Americans" means residents of any nation from Canada through Latin America, but our language fails to provide us with a more concise adjective such as "Canadian" or "Mexican."

2. Recent research has shown that communication is more effective when based on an understanding of preexisting mental models, even in comparison with communication designed by experts familiar with general communication strategies (Morgan, et al. 1992).

3. Marketing Intelligence Service reported that for new products, environmental claims on packages doubled from 1989 to 1990. An audit by J. Walter Thompson, an advertising agency, found that the number of environmental claims in advertisements quadrupled during 1990 (*Consumer Reports* 1991), although by 1992 they had begun to drop off (*Consumer Reports* 1992).

4. Germany has had a government-approved environmental labeling program since 1978, and German companies report a hefty 10 percent to 30 percent jump in sales after products receive the label. The program is now having a substantial effect on the way products are designed and manufactured (Simons 1993).

5. Such trends may be true for other postmaterialist values, but the argument about environmentalism emerging only in industrial or postindustrial societies fails a simple test: public environmental sentiment is as strong among developing nations as industrialized ones (Brechin and Kempton 1994; Dunlap, Gallup, and Gallup 1992, 1993; Louis Harris 1989). Of course, environmental movements may take radically different forms in developing countries (Guha 1990), and developing country governments may have different agendas in international environmental accords (Grubb et al. 1993, 26–37).

6. Sometimes such models are contrasted with those constructed deliberately by the scientific community, by calling them "naive models" (McCloskey, Cara-

mazza, and Green 1980) or "folk models" (Kempton 1987). The former term, derived from studies in an educational context, makes an unwarranted presumption of dysfunctionality. The term *folk model,* in suggesting oral transmission and wide distribution outside of specialists, is closer to what we mean by a cultural model.

7. For example, Marshall (1957) reports that !Kung Bushmen were sure that inherent properties of hunters and their game would combine, at the time of the kill, to affect the weather that day. Marshall could not determine any function for this belief, nor any harm caused by it.

Chapter 2

1. Garro (1988) and Holland and Skinner (1987) are among the exceptions to this tendency.

2. One antienvironmental group we know of is called the "wise use" movement (Hennelly 1992) or "property rights movement" (Schneider 1992). This includes several groups, but it is difficult to determine the extent to which they are truly ideologically oriented membership groups versus a mouthpiece for existing financial interests. For example, one such organization is the Western States Public Land Coalition (in Pueblo, Colorado), which publishes a newsletter "People for the West!" and bills itself as a "grassroots movement." But twelve of the thirteen members of the board of directors are mining industry executives. The newsletter articles cover logging, mining, grazing, and recreational off-road vehicles. Articles itemize reasons the U.S. Forest Service should not charge more for mining and logging rights, but there is no discussion of ideological issues of humanity and the environment. At a local level, Priscilla Weeks has documented occupationally organized local groups fighting environmental regulations that affect their livelihood. More commonly, ideologically driven antienvironmental positions are promoted in service to a broader political ideology (e.g., Limbaugh 1993).

3. In quotations, ellipses (. . .) indicate material we have deleted, and brackets [ ] indicate our postinterview additions of clarification or paraphrase. See appendix A for more on our transcript notation.

4. We use the term *ultraviolet* in the text to refer to UV-B, the midpart of the UV spectrum (280–320 nm). This range of UV is highly damaging to life, and more of it is now reaching earth's surface due to ozone depletion.

5. The figure of 27 million deaths applies to people born worldwide up to 2075, the approximate residence time of the CFCs. The usually fatal form of skin cancer is called melanoma. When one also includes nonmelanoma carcinomas, the more common and treatable forms, a total of 132 million skin cancers are estimated to be averted by this treaty (Shabecoff 1987). Some damage to the ozone layer has already occurred. The U.S. EPA estimated in April 1991 that winter ozone levels declined by 5 percent over the United States during the

eighties. From the ozone loss that has already occurred, they estimate that in the United States alone, this will cause 12 million skin cancers over fifty years, over 200,000 of which will be fatal (Stevens 1991, A1). These figures are probably conservative, as subsequent measurements show greater ozone loss at high latitudes, and in summer, than anticipated in the EPA's calculations (Appenzeller 1991, 645).

6. The comparative figures for the world wars include combatants and noncombatants, in all countries. These estimates are drawn from Keegan (1989) and *Encyclopedia Britannica* (1991).

7. Secretary Hodel's suggestion was reported in the 30 May 1987 issues of the *New York Times* and the *Los Angeles Times*. For one set of follow-up stories, see the 6 July, 3 August, and 10 August 1987 issues of *The New Republic*.

8. Strictly speaking, the mechanism of a greenhouse building is different, since its elevated temperatures are due more to the glass structure's trapping of warm air than reflecting of heat. The analogy is still helpful.

9. As an example of the debates around specific numbers, the table gives the role of gases in "radiative forcing." However, for CFCs, their role in climate change may be much less than the 24 percent figure would suggest. Recent findings suggest that stratospheric ozone depletion (caused by CFCs) cools the lower atmosphere, and the cooling may be more than enough to offset the increased greenhouse effect of CFCs. So even though CFCs are a greenhouse gas, and even though the CFC ban is clearly desirable to preserve earth's protective ozone layer, banning CFCs may not help at all in reducing global climate change (IPCC 1992, 20).

10. The question of whether or not these effects can already be measured was always difficult, due to the large natural fluctuations in weather. Unfortunately, this question got more complicated in 1991 when Mt. Pinatubo (in the Philippines) erupted, spewing ash and sulfur dioxide into the stratosphere. These substances are expected to keep the earth cooler than it would otherwise be through 1994 or 1995, enough cooler to mask any effect of global warming during these years (IPCC 1992, 8; Kerr 1992).

11. A thorough discussion of adaptation strategies can be found in a National Academy of Sciences panel report (NAS 1991b). That NAS panel did not argue that adaptation should be a substitute for prevention, but rather that costs from global warming damage would be relatively small and that adaptation will probably be required even with prevention. For a response to the adaptation argument, see Mathews (NAS 1991c, 45–46).

Chapter 3

1. Our interview questions deliberately call the natural world by three different names to emphasize the perspective of each question. We ask about *using natural resources*, *protecting* the *environment*, and *a relationship with nature*. In this

chapter's analytic discussion, we will similarly select among these terms to highlight which aspect of the natural world we are discussing at the time.

2. *Italics* are used for our interview questions, especially in quotations when we need to distinguish our questions from informant responses.

3. In the semistructured interviews, none of our informants used "spaceship" as a metaphor (Dunlap's question wording was inspired by writings of environmentalists), and many of our informants were more specific than these survey questions about the closed nature of flows of waste. These findings suggest some possible rewording to make these questions better fit the way that laypeople think about the concepts.

4. In the short term, hunters can be seen as accepting with grace their place in the food chain, and, by acting as a part of nature, helping to preserve the natural balance. A longer-term view, not mentioned by any of these informants discussing hunting, is that the reason that humans must play this role is that they have eliminated competing predators, in this case wolves.

5. Although we feel that evidence for an American model of species interdependencies and nonintervention is overwhelming, we note here that two of our survey statements—47 and 125—could be interpreted as counterevidence. Statement 47 was "The way to solve environmental problems is to stop human activities that disturb the environment, and let nature heal itself, rather than actively intervening to 'fix' the environment." Statement 125 was "If we had the technology to change climate for the better, to improve the human condition, we should do so." A noninterventionist position would be to agree with statement 47 and disagree with 125. Apart from Earth First! members, majority responses did not meet our expectations (see appendix C for percentages). Unfortunately, our wording in both these questions was ambiguous. People may have disagreed with 47 as a way of saying that society should "actively" protect the environment. Similarly, they may have agreed with 125 because they felt we should use technologies like renewable energy to prevent climate change (the question was intended to refer to deliberate climate modification). The nonintervention model should be tested further with more refined survey questions. We nevertheless feel safe considering it a majority model based on our other survey questions, on the semistructured interviews, and on the Dunlap, Gallup, and Gallup data.

6. This "Indian model" may refer to a tradition of the Iroquois Confederacy. It has gained recognition recently because it is cited prominently by a mail-order company that specializes in environmental products.

7. We use the word *myth* to refer to a story passed from person to person which, regardless of its objective truth, captures something culturally or symbolically significant (Lévi-Strauss 1966).

8. We agree with Aldo Leopold that modern societies have in some ways improved upon the values of the past:

For one species to mourn the death of another is a new thing under the sun. The Cro-Magnon who slew the last mammoth thought only of steaks. The sportsman who shot the last [carrier] pigeon thought only of his prowess. The sailor who clubbed the last auk thought of nothing at all. But we, who have lost our pigeons, mourn the loss. Had the funeral been ours, the pigeons would hardly have mourned us. In this fact, rather than in Mr. DuPont's nylons or Mr. Vannevar Bush's bombs, lies objective evidence of our superiority over the beasts. (Leopold 1970, 117)

Chapter 4

1. Sixteen recognized the specific term *greenhouse effect.* A seventeenth did not recognize that term but had heard that, as he put it, "in the twenty-first century it's going to be unbearably warm."

2. News article on page 23 of the 28 May 1990 issue.

3. For example, in folk biology, see Berlin (1972), or in religion, see Lessa and Vogt (1972).

4. In fact, a scientist tried to correct these facts twenty-five years ago (Broecker 1970), criticizing as misleading the then publicly expressed concern about running out of oxygen due to pollution. The fact that the misconception is a majority position today, a quarter century later, supports our contention that this "correct the numbers" strategy is insufficient.

5. CO_2 is also removed via ocean sedimentation trapping the calcium carbonate in dead organisms' shells. This process is reversed by volcanic outgassing of CO_2, a return process generally assumed to be in equilibrium with sedimentation.

6. There has been some analysis showing higher human mortality due to higher temperatures (Kalkstein and Davis 1989; Kalkstein 1991), but higher mortality is associated with long, unseasonably hot weather conditions rather than regions that have hotter overall temperatures.

7. This is an argument for referring to the anticipated changes as "global climate change," which is really more the point than "global warming," per se. In fact, we tried using "global climate change" in drafts of this book, but found it wordy and found that some readers were confusing global environmental change with global climate change. By now "global warming" seems to have gained mainstream acceptance, so we have adopted it here.

8. In the semistructured interviews with laypeople, as many people mentioned human causes as mentioned natural causes (Kempton 1991a, 191). In part, the pollution answer may have been enhanced because we initially described the interview topic as "weather and the environment." There is also an unintended ambiguity in the question word "affect," as indicated by one informant's paraphrase of our question as "things that could change the natural weather patterns." Other data described subsequently in this section—see our survey statements 17 and 135 and Farhar's (1976) data—show that the plausibility of human effects on weather is not a finding limited to our semistructured interviews.

9. Kempton notes that this follow-up comment violates his interviewing ethics by stating something he did not believe to be true. In the transcript, he said "*there could be something to that*" because he was dumbfounded, yet needed to respond quickly and positively to encourage the informant to expand on her "private theory."

Chapter 5

1. Question I.J, in the semistructured interview protocol in appendix C, did ask about children. However, this question was added after interview number 39, almost at the end of the semistructured interviews. All discussions of descendants in this section's quotations were initiated by the informants themselves.

2. Since Earth First! members are somewhat younger than members of the other groups (median age thirty; see appendix B), fewer may actually have families to feed. (We do not have direct data on their family or marital status.) Nonetheless, we feel that such demographic differences are less likely to explain the large group differences in answers to question 63 on "feeding my family" than might their environmental commitment.

3. The economic concept of discounting future costs can be explained by analogy to compound interest paid by a bank. If you put $100 in a savings account today at 10 percent interest, you will have $110 in one year and $121 in two years. By this reasoning, $121 two years in the future is worth the same as $100 today. This is the concept of discounting the future: that when one makes decisions about the future, whether considering benefits or costs, they must be discounted to their values today in order to compare them with costs or benefits incurred today.

4. Jeremy Bentham, the founder of modern utilitarianism, calculated the value of a thing by the pleasure and pain it gave to all sentient beings, not just humans. Here we use the term *utilitarianism* to refer only to anthropocentric utilitarianism.

5. Despite the high agreement by respondents, it could be argued that the premise of this question is scientifically incorrect: the greatest climate change would occur in higher latitudes, while the vast majority of species live in tropical forests. The greatest global threat to species is not global warming but loss of tropical forest habitat, principally due to nonsustainable timber harvests and conversion of land to agricultural or other use. Nevertheless, in the ecological zones inhabited by most American informants, climate change (in conjunction with other human changes) is a threat to some species.

6. Upon reading an earlier draft of part of this discussion, an entomologist wrote to report a similar bias among biologists: "There is even a problem among scientists. Most of the scientists have focused their concern on the large, beautiful animals and plants and little concern is directed toward the microscopic organisms and other small organisms, like insects. Yet the structure and functioning

of the ecosystem is dominated by the little things" (David Pimentel, Cornell University, letter of 12 September 1990).

7. In an often-cited comment on Heberlein, Dunlap and Van Liere (1977) point out that his research includes only situations in which environmental destruction will affect other humans. Thus, they argue, Heberlein's data could be explained without a land ethic or biocentric norms. Instead, they can simply be explained as another application of the interpersonal golden rule ("Do unto others as you would have them do unto you"). Now, two decades after Heberlein's study, our data suggest that a biocentric value has emerged more clearly and unambiguously. Dunlap, a skeptic in 1977, now reaches a similar conclusion: "Over time, as the ecological worldview has evolved and spread, the ethical implications of it have also become stronger. I would argue, for example, that a true land ethic (concerned with the effects of our actions on non-humans) is far stronger now than it was in the mid-seventies. At the same time, I would also argue that for most people, concern for human consequences is still a more powerful motivation for pro-environmental behaviors than is concern for non-human consequences." (memorandum from Dunlap to Kempton, 15 January 1992).

8. In the previous section, we quoted Ronald saying "The Creator intended that nature be used by man," and "I don't think the environment is a separate entity that has any kind of right to exist." Is this contradictory with his "part of nature" and "harmony" statement? We would not be surprised to find an informant self-contradictory occasionally (we probably are ourselves, at least once or twice in this book). However, we think something more interesting is going on here. There are many cultural models, beliefs, and values that support a proenvironmental stance. We will demonstrate more clearly in chapter 8 that even those people like Ronald who could be called "antienvironmental" share many of the perspectives of the proenvironmental view.

9. The context of this quotation in the full transcript makes it clear that Gerard was using "our species" to refer not to humans but to the other plant and animal species with whom we share our planet.

10. Based on a few passages such as Wilbur's in the initial open-ended interviews, Kempton (1991a) concluded that an abstract species preservation ethic was not strongly felt. Now, with the complete set of open-ended interviews and the survey, that conclusion seems to have been premature.

11. At the other end of the spectrum of causality would be the "biophilia hypothesis," which posits that environmental values are innate to humans and carry adaptive advantage (Wilson and Kellert 1993). Even if some environmental values are innate, they clearly vary greatly among people and among cultures. For example, we doubt that environmental values were as closely tied to other core American values during the nineteenth century.

Chapter 6

1. These policies for global warming may also improve problems of air pollution, foreign exchange for imported petroleum, and so forth, but we did not

discuss these aspects in the interviews, and they were rarely brought up by informants during policy discussions.

2. A survey in the United Kingdom required respondents to pick just one or two "most effective" ways to reduce CO_2. Catalytic converters, fewer cars on the road, and recycling topped their list. Policies that analysts would typically recommend, "better insulation" and "use less electricity," were chosen by far fewer respondents (MORI 1990). After reading the U.K. study, we added recycling to our survey's list of possible policies to reduce global warming. Recycling was picked most frequently of the twelve measures on our list. We do not discuss the results from our policy list in detail because our question asked for "good ways to effectively prevent" global warming, which was ambiguous as to whether it was eliciting preference or effectiveness. Read et al. suggested that their respondents also picked solutions partly on the basis of what was "environmentally good." Conversely, an earlier study found that many people incorrectly identified "environmentally bad" things, such as pesticides or nuclear power, as causes of global warming (Doble, Richardson, and Danks 1990, 1).

3. Two energy sources, nuclear energy and renewable energy, were asked about either in the semistructured interview or the survey, but we do not discuss them here because the data were not sufficient for an analysis that would go beyond the existing risk perception and public opinion literature on these subjects.

4. Some analysts use the term *limitation* for what we call *prevention*. The argument for using "limitation" is that the greenhouse gases already released commit the planet to some climate change, so the best that can be done is to limit the change from going a lot further; it is impossible to totally prevent any change from happening. Although this is true, we use "prevention" because it more clearly makes the contrast with "adaptation."

5. At 30 percent efficiency of the spray gun times 20 percent of the solids actually being applied, we calculate 6 percent rather than 7 percent. Perhaps the 30 percent and 20 percent figures he gave us were rounded off.

6. Calling a 100 percent energy tax extreme is very much a U.S. perspective; European gasoline taxes range from 100 percent to 300 percent. Our proposed tax was computed on the retail cost of three fuels for simplicity in explanation, but a tax on carbon content (Chandler and Nicholls 1990) may be a sounder policy for energy efficiency to reduce global warming. The Clinton Btu tax proposal was based approximately on carbon levels, but adjusted (substantially) to equalize regional impacts. We judged that explaining such refinements over price-based energy taxes would cost interview time with little gain in value of the data.

7. From an advertisement with the headline "Take a shower, pay a tax," run by the Affordable Energy Alliance; this example from *The New Republic,* inside back cover, 28 June 1993.

Chapter 7

1. For comparison, case studies of nonspecialists are presented in appendix D.

2. In our view, mangoes in North Dakota is a huge exaggeration, given expected temperature increases even for sixfold CO_2 increases.

3. The list is within II.E in the semistructured interview protocol, in appendix C.

4. Alvin's appeal to common sense is not valid in terms of physical science. The global heat contributed from combustion is vanishingly small in comparison to the solar radiative forcing effect of the CO_2, which is a combustion by-product.

5. Currently, all environmental compliance costs combined are about 1.5 percent of U.S. gross national product (OTA 1991:10). Gerard may be thinking of industry projections of increased automobile costs under the Clean Air Act of 1990, added to incremental costs of all prior environmental regulations. This would be closer to the 20 percent example figure.

6. He confused only a few aspects of the science, for example, that greenhouse gases "prevent circulation of currents" in the atmosphere, and none of the details confused were of policy significance.

7. In explaining his skepticism of models, he noted the spectacular failure of the Congressional Budget Office's estimate of catastrophic health care costs as being low by a factor of ten, which resulted in the legislation being canceled the year after it was passed.

8. There are some confusions on the underlying science, such as "carbon dioxide is diminishing the ozone layer that keeps the sun out," and CO_2 was the only greenhouse gas mentioned. These had no significant impact on his interpretation of proposals or strategies for his own reaction to the problem.

Chapter 8

1. All individuals taking the survey were analyzed together as one group to determine the axes and each individual's position on the graphs. The figures in this chapter display each group separately because the graphs are more easily read that way.

2. We have simplified Milbrath's figure by reducing the number of groups and by plotting environmentalism as the main axis, eliminating his axis for social activism.

3. It is difficult to determine the precise point marking the ending of environmental beliefs and the beginning of other beliefs such as the need for basic social change or the limits to society's growth. Upon reading the passage associated with this note, Lester Milbrath remains convinced that these issues are an integral part of the environmental paradigm. He argues: "Just because concepts and issues are not brought up when you ask questions in an open-ended way does

not mean that the concepts or issues do not exist in their minds, or that they don't have strong beliefs about them, nor even that people do not see connections to environmental protection. . . . Given a moment's thought, most people could readily see the connection." (letter of 23 March 1994).

4. Which statements involve value trade-offs and which do not? We would judge the following statements to involve some form of value trade-offs: 24, 25, 28, 29, 66, 75, 80, 112, 115, 129, 144, 145, 147.

Chapter 9

1. Reported at the workshop Methodological Issues in Global Modelling, Universidad Nacional Autónoma de México, November 1990.

Appendix A

1. The semistructured interview protocols were not fully comparable between the two pretest interviews and the remainder of the sample. We sometimes use them for quotations when the data are comparable, but they are not counted when we make quantitative statements about how many people in the semistructured interviews made some specific point.

Appendix C

1. Stephen H. Schneider, "The Greenhouse Effect: Science and Policy," *Science* 243 (10 February 1989):771–781. He estimates 2 to 6°C, that is, 3.6 to 10.8°F. See also "Briefing," *Science* 248 (25 May 1990):962. Sixty-five percent of 330 polled "global env. change scientists" believe 2°C warming better than fifty-fifty probability. Ninety percent favored taking immediate steps to reduce CO_2 emissions.

2. James G. Titus, "The Causes and Effects of Sea Level Rise," in *The Challenge of Global Warming*, ed. Dean Edwin Abrahamson, (Washington, D.C.: Island Press), 169, 188. P. 169 cites a 36–212 cm rise by 2075, that is, from 1.2 ft to 6.9 ft; p. 188 notes that a 30 cm rise erodes beaches 20 to 60 m).

3. Leslie Roberts, "How Fast Can Trees Migrate?" *Science* 243 (10 February 1989):735–737. Beech is the most displaced of four trees studied.

4. Leslie Roberts, "Is There Life after Climate Change?" *Science* 242 (18 November 1989):1010–1012.

5. U.S. EPA, "Policy Options for Stabilizing Global Climate," Draft Report to Congress, February 1989, I-11, fig. 1.2.

6. *Ibid.*, fig. 1.2. The 75 percent cut holds atmospheric CO_2 approximately level; there is some lag in the temperature and other effects.

Appendix D

1. In such an intensive question sequence, it is important to ask whether we led the informant to those answers. Most of these reported values seem reliable because the questioning was nondirective, and the informant readily elaborated once she acknowledged her underlying values. We cannot make as strong a case for her "Christian" response, since it followed a more explicitly leading interviewer prompt, and she provided no further elaboration (see quotation). Another methodological issue is that informants sometimes give the normatively correct answer in an interview but act differently in the privacy of the voting booth or marketplace. Since Tara has not yet mailed in the environmental group applications waiting on the table, her values have not yet motivated her to act.

References

Abrahamson, Dean. 1992. "Climatic Change and Energy Supply: A Comparison of Solar and Nuclear Options." Pp. 115–140 in *Energy and Environment: The Policy Challenge,* ed. John Byrne and Daniel Rich. Vol. 6 Energy Policy Studies series. New Brunswick, N.J.: Transaction Publishers.

Agar, Michael H. 1980. *The Professional Stranger: An Informal Introduction to Ethnography.* New York: Academic Press.

Anderson, Atholl. 1990. *Moas and Moa-Hunting in Prehistoric New Zealand.* New York: Cambridge University Press.

Appenzeller, Tim. 1991. "Ozone Loss Hits Us Where We Live." *Science* 254(1 November):645.

Beauchamp, Tom L. 1982. *Philosophical Ethics: An Introduction to Moral Philosophy.* New York: McGraw-Hill.

Benedick, Richard Elliot. 1991. *Ozone Diplomacy: New Directions in Safeguarding the Planet.* Cambridge, Mass.: Harvard University Press.

Berke, Richard L. 1990. "Oratory of Environmentalism Becomes the Sound of Politics." *New York Times,* 17 April, A1.

Berlin, Brent. 1972. "Speculations on the Growth of Ethnobotanical Nomenclature." *Language in Society* 1(1):51–86. Reprinted 1977 in *Sociocultural Dimensions of Language Change,* ed. Ben G. Blount and Mary Sanches, pp. 63–102. New York: Academic Press.

Bernard, H. Russell. 1994. *Research Methods in Cultural Anthropology.* 2d ed. Newbury Park, Calif.: Sage Publications.

Berry, Thomas. 1988. *The Dream of the Earth.* San Francisco: Sierra Club Books.

Boster, James, and Jeffrey C. Johnson. 1989. "Form or Function: A Comparison of Expert and Novice Judgments of Similarity among Fish." *American Anthropologist* 91(4):866–889.

Boster, James, and Susan Weller. 1990. "Cognitive and Contextual Variation in Hot–Cold Classification." *American Anthropologist* 92:171–179.

Brechin, Steven R., and Willett Kempton. 1994. "Global Environmentalism: A Challenge to the Postmaterialism Thesis?" *Social Science Quarterly* 75(2):245–269.

Broecker, Wallace S. 1970. "Man's Oxygen Reserves." *Science* 168(26 June): 1537–1538.

Bryant, Bunyan, and Paul Mohai, eds. 1992. *Race and the Incidence of Environmental Hazards: A Time for Discourse.* Boulder, Col.: Westview Press.

Bullard, Robert D. 1990. *Dumping in Dixie: Race, Class and Environmental Quality.* Boulder, Col.: Westview Press.

Bundestag. 1989. *Protecting the Earth's Atmosphere: An International Challenge.* Interim Report of the Study Commission of the 11th German Bundestag "Preventive Measures to Protect the Earth's Atmosphere." Bonn: German Bundestag, Publ. Sect.

Burch, William R., Jr. 1971. *Daydreams and Nightmares: A Sociological Essay on the American Environment.* New York: Harper and Row.

Buttel, Frederick H., and Peter J. Taylor. 1992. "Environmental Sociology and Global Environmental Change: A Critical Assessment." *Society and Natural Resources* 5:211–230.

Byrne, John, Constantine Hadjilambrinos, and Subodh Wagle. 1994. "Distributing Costs of Global Climate Change." *IEEE Technology and Society Magazine* (Spring):17–24, 32.

Caldeira, Ken, and James F. Kasting. 1993. "Insensitivity of Global Warming Potentials to Carbon Dioxide Emission Scenarios." *Nature* 366(18 November):251–253.

Carson, Rachel. 1962. *Silent Spring,* Boston: Houghton Mifflin.

Cess, R. D. et al. 1993. "Uncertainties in Carbon Dioxide Radiative Forcing in Atmospheric General Circulation Models." *Science* 262(19 November):1252–1255.

Chandler, William U., and Andrew K. Nicholls. 1990. "Assessing Carbon Emissions Control Strategies: A Carbon Tax or A Gasoline Tax?" ACEEE Policy Paper No. 3. Washington, D.C.: American Council for an Energy-Efficient Economy.

Chernela, Janet M. 1982. "Indigenous Forest and Fish Management in the Uaupes Basin of Brazil." *Cultural Survival Quarterly* 6(2):17–18.

Chernela, Janet M. 1985. "Indigenous Fishing in the Neotropics: The Tukanoan Uanano of the Blackwater Uaupes River Basin in Brazil and Colombia." *Interciencia* 10(2):78–86.

Chernela, Janet. 1987. "Endangered Ideologies: Tukano Fishing Taboos." *Cultural Survival Quarterly* 11(2):50–52.

Childs, I. R. W., A. Auliciems, T. J. Hundloe, and G. T. McDonald. 1988. "Socio-economic Impacts of Climate Change: Potential for Decision-Making in

Redcliffe, Queensland." Pp 648–679 *Greenhouse Planning for Climate Change*, ed. G. I. Pearman. Leiden: E. J. Brill.

Church, George J. 1993. "A Call to Arms." *Time* (22 February):27–30.

Clark, William C., ed. 1982. *Carbon Dioxide Review*. Oxford, U.K.: Clarendon Press.

Clement, John. 1982. "Students' Preconceptions in Introductory Mechanics." *American Journal of Physics* 50(1):66–71.

Clements, Frederic L. 1916. *Plant Succession*. Washington, D.C.: Carnegie Institution.

Cline, William R. 1992. *Global Warming: The Economic Stakes*. Washington, D.C.: Institute for International Economics.

Collins, H. 1987. "Certainty and the Public Understanding of Science." *Social Studies of Science* 17(4):689–713.

Consumer Reports. 1991. "Selling Green," 56(10):687–692.

Consumer Reports. 1992. "Governing the Green Marketplace," 57(11):707.

Cotgrove, Stephen F. 1982. *Catastrophe or Cornucopia: The Environment, Politics and the Future*. Chichester, N.Y.: Wiley.

Council of Economic Advisors. 1990. "Economic Report of the President, with Annual report of the Council of Economic Advisors." (See section on Global Environmental Issues, February.)

Craig, Paul P., Harold Glasser, and Willett Kempton. 1993. "Ethics and Values in Environmental Policy: The Said and the UNCED." *Environmental Values* 2:137–157.

Cronon, William. 1991. *Nature's Metropolis: Chicago and the Great West*. New York: W. W. Norton.

Crosby, Alfred W. 1986. *Ecological Imperialism: The Biological Expansion of Europe, 900–1900*. Cambridge, U.K.: Cambridge University Press.

DeCicco, John, James Cook, Dorene Bolze, and Jan Beyea. 1990. *CO_2 Diet for a Greenhouse Planet: A Citizen's Guide to Slowing Global Warming*. New York: National Audubon Society.

DeCicco, John, and Marc Ross. 1993. "An Updated Assessment of the Near-term Potential for Improving Automotive Fuel Economy." Report, 99 pp. Washington, D.C.: American Council for an Energy-Efficient Economy.

Devall, Bill. 1991. "Deep Ecology and Radical Environmentalism." *Society and Natural Resources* 4(3):247–258.

Devall, Bill, and George Sessions. 1985. *Deep Ecology: Living as if Nature Mattered*. Layton, Utah: Peregrine Smith Books.

De Young, Raymond. 1991. "Some Psychological Aspects of Living Lightly: Desired Lifestyle Patterns and Conservation Behavior." *Journal of Environmental Systems* 20(3):215–227.

Diamond, Jared M. 1986. "The Environmentalist Myth." *Nature* 324(6 November):19–20.

Doble, John, Amy Richardson, and Allen Danks. 1990. *Science and the Public, Volume III: Global Warming Caused by the Greenhouse Effect*. New York, N.Y.: The Public Agenda Foundation.

Douglas, Mary, and Aaron Wildavsky. 1982. *Risk and Culture: An Essay on the Selection of Technical and Environmental Dangers*. Berkeley: University of California Press.

Dunlap, Riley E., George H. Gallup, Jr., and Alec M. Gallup. 1992. "The Health of the Planet Survey: A Preliminary Report on Attitudes on the Environment and Economic Growth Measured by Surveys of Citizens in 22 Nations to Date." A George H. Gallup Memorial Survey. Princeton, N.J.: The George Gallup International Institute.

Dunlap, Riley E., George H. Gallup, Jr., and Alec M. Gallup. 1993. "Of Global Concern: Results of the Health of the Planet Survey." *Environment* 35(9):7–15, 33–39.

Dunlap, Riley E., and Rik Scarce. 1991. "The Polls—Poll Trends: Environmental Problems and Protection." *Public Opinion Quarterly* 55:713–734.

Dunlap, Riley E., and Kent D. Van Liere. 1977. "Land Ethic or Golden Rule: Comment on 'Land Ethic Realized' by Thomas A. Heberlein." *Journal of Social Issues* 33(3):200–207.

Dunlap, Riley E., and Kent D. Van Liere. 1978. "The 'New Environmental Paradigm': A Proposed Measuring Instrument and Preliminary Results." *Journal of Environmental Education* 9:10–19.

Edgerton, Robert B. 1992. *Sick Societies: Challenging the Myth of Primitive Harmony*. New York: Free Press.

Egan, Timothy. 1992. "Chief's 1854 Lament Linked to Ecological Script of 1971," *New York Times*, 21 April, A1, A17.

Erlich, P. 1968. *The Population Bomb*. New York: Ballantine.

Farhar, Barbara C. 1976. "The Impact of the Rapid City Flood on Public Opinion about Weather Modification." *Pacific Sociological Review* 19(1):117–144.

Farhar, Barbara C. 1977. The Public Decides about Weather Modification." *Environment and Behavior* 9(3):279–310.

Farhar, Barbara C. 1993. "Trends in Public Perceptions and Preferences on Energy and Environmental Policy." National Renewable Energy Laboratory Report No. NREL/TP-461-4875, 419 pp. Springfield, Va.: National Technical Information Service. (Abbreviated version to appear in 1994 *Annual Review of Energy and the Environment*.)

Farhar, Barbara C., Jack A. Clark, Lynn A. Sherretz, Jerry Horton, and Sigmund Krane. 1979. "Social Impacts of the St. Louis Urban Weather Anomaly." Final

Report, Vol. 2. Boulder, Col.: Institute of Behavioral Science, University of Colorado.

Flavin, Christopher. 1992. "Building a Bridge to Sustainable Energy." Pp. 27–45 in *State of the World 1992*, ed. Lester R. Brown. New York: W. W. Norton.

Garro, Linda. 1988. "Explaining High Blood Pressure: Variation in Knowledge about Illness." *American Ethnologist* 15: 98–119.

Geisel, Theodor Seuss (Dr. Seuss). 1971. *The Lorax*. New York: Random House.

Gentner, Dedre, and Albert L. Stevens. 1983. *Mental Models*. Hillsdale, N.J.: Lawrence Erlbaum Associates.

Gore, Al. 1992. *Earth in the Balance: Ecology and the Human Spirit*. Boston: Houghton Mifflin.

Greenberg-Lake and Tarrance Group. 1993. A Sustainable Energy Blueprint: National Opinion Poll. Report available from Alliance to Save Energy and Friends of the Earth, Washington, D.C.

Greene, David L. 1987. *Research Priorities in Transportation and Energy*. Transportation Research Circular No. 323, September. Washington, D.C.: Transportation Research Board, National Research Council.

Gross, Daniel. 1990. "Seed Money: A Billion Points of Shade?" *The New Republic*, 21 May, 11–12.

Grubb, Michael, Matthias Koch, Abby Munson, Francis Sullivan, and Koy Thomson. 1993. *The Earth Summit Agreements: A Guide and Assessment*. London: EarthScan Publications and Royal Institute of International Affairs.

Guha, Ramachandra. 1990. *The Unquiet Woods: Ecological Change and Peasant Resistance in the Himalayas*. Berkeley: University of California Press.

Harman, Willis W. 1977. "The Coming Transformation." *The Futurist* 11:106–112.

Harte, John, and Robert H. Socolow. 1971. *Patient Earth*. New York: Holt, Rinehart and Winston.

Hays, Samuel P. 1987. *Beauty, Health, and Permanence: Environmental Politics in the United States, 1955–1985* (in collaboration with Barbara D. Hays). New York: Cambridge University Press.

Hays, Samuel P. 1992. "Environmental Political Culture and Environmental Political Development: An Analysis of Legislative Voting, 1971–1989." *Environmental History Review* 16(2):1–22.

Heberlein, Thomas A. 1972. "The Land Ethic Realized: Some Social Psychological Explanations for Changing Environmental Attitudes." *Journal of Social Issues* 28(4):79–87.

Henderson, H. 1976. "Ideologies, Paradigms and Myths: Changes in our Operative Social Values." *Liberal Education* 62:143–157.

Hennelly, Robert. 1992. "Getting Wise to the 'Wise Use' Guys." *The Amicus Journal* (Fall):35–38.

Holden, Constance. 1990. "Climate Experts Say It Again: Greenhouse Is Real" *Science* 248(25 May):964.

Holland, Dorothy, and Naomi Quinn. 1987. *Cultural Models in Language and Thought*. London: Cambridge University Press.

Holland, Dorothy, and Debra Skinner. 1987. "Prestige and Intimacy: The Cultural Models behind Americans' Talk about Gender Types." Pp. 87–111 in *Cultural Models in Language and Thought,* ed. Dorothy Holland and Naomi Quinn. London: Cambridge University Press.

Holmes, Richard T., and Thomas W. Sherry. 1992. "Site Fidelity of Migratory Warblers in Temperate Breeding and Neotropical Wintering Areas: Implications for Population Dynamics, Habitat Selection, and Conservation." Pp. 563–578 in *Ecology and Conservation of Neotropical Migrant Landbirds* ed. John M. Hagan III and David W. Johnston. Washington, D.C.: Smithsonian Institution Press.

Houghton, R. A., and David L. Skole. 1990. "Carbon" Pp. 393–408 in *The Earth as Transformed by Human Action: Global and Regional Changes in the Biosphere over the Past 300 Years,* ed. B. L. Turner et al. Cambridge, U.K.: Cambridge University Press.

Hubert, Lawrence, and James Schultz. 1976. "Quadratic Assignment as a General Data Analysis Strategy." *British Journal of Mathematical and Statistical Psychology* 29:190–241.

Inglehart, Ronald. 1977. *The Silent Revolution: Changing Values and Political Styles among Western Publics.* Princeton, N.J.: Princeton University Press.

Intergovernmental Panel on Climate Change (IPCC). 1990. *Climate Change: The IPCC Scientific Assessment,* ed. John T. Houghton, G. J. Jenkins, and J. J. Ephraums. Cambridge, U.K.: Cambridge University Press.

Intergovernmental Panel on Climate Change (IPCC). 1992. *The 1992 IPCC Supplement.* Manuscript, February, 71 pp. (Available from World Meterological Organization/United Nations Environmental Program.)

Johansson, Thomas B., Henry Kelly, Amulya K. N. Reddy, and Robert H. Williams, eds. 1993. *Renewable Energy: Sources for Fuels and Electricity.* Washington, D.C.: Island Press.

Johnson, Allen. 1978. "In Search of the Affluent Society." *Human Nature* (September):50–59.

Johnson-Laird, P. N. 1980. "Mental Models in Cognitive Science." *Cognitive Science* 4:71–115.

Kalkstein, Laurence S. 1991. "A New Approach to Evaluate the Impact of Climate on Human Mortality." *Environmental Health Perspectives* 96:145–150.

Kalkstein, Laurence S., and Robert E. Davis. 1989. "Weather and Human Mortality: An Evaluation of Demographic and Interregional Responses in the United States." *Annals of the Association of American Geographers* 79(1):44–64.

Keegan, John, ed. 1989. *Times Atlas of the Second World War.* New York: New York Times Books.

Kempton, Willett. 1987. "Two Theories of Home Heat Control." Pp. 222–242 in *Cultural Models in Language and Thought,* ed. Dorothy Holland and Naomi Quinn. Cambridge, U.K.: Cambridge University Press.

Kempton, Willett. 1991a. "Lay Perspectives on Global Climate Change." *Global Environmental Change: Human and Policy Dimensions* 1:183–208. Reprinted in *Energy Efficiency and the Environment: Forging the Link,* E. Vine, D. Crawley, and P. Centolella, eds., pp. 29–69. Washington, D.C.: American Council for an Energy-Efficient Economy.

Kempton, Willett. 1991b. "Public Understanding of Global Warming." *Society and Natural Resources* 4:331–345.

Kempton, Willett. 1993. "Will Public Environmental Concern Lead to Action on Global Warming?" *Annual Review of Energy and the Environment* 18:217–245.

Kempton, Willett, and Paul P. Craig. 1993. "European Perspectives on Global Climate Change." *Environment* 35(3):16–20, 41–45.

Kempton, Willett, C. K. Harris, J. G. Keith, and J. S. Weihl. 1982. "Do Consumers Know 'What Works' in Energy Conservation?" Pp. 429–438 in *What Works: Documenting Energy Conservation in Buildings.* J. Harris and C. Blumstein, eds. Washington, D.C.: American Council for an Energy-Efficient Economy.

Kerr, J. B., and C. T. McElroy. 1993. "Evidence for Large Upward Trends of Ultraviolet-B Radiation Linked to Ozone Depletion." *Science* 262(5136):1032–1034.

Kerr, Richard A. 1992. "1991: Warmth, Chill May Follow." *Science* 255:281.

Kimble, George H. T. 1962. "But Somebody Does Something about It." *New York Times Magazine,* 8 July, 11ff.

Koppel, Ted. 1988. "The Greenhouse Effect." *ABC News Nightline.* Broadcast 7 September. (Transcript available from Journal Graphics, New York, N.Y.)

Kuhn, Thomas. 1962. *The Structure of Scientific Revolutions.* Chicago: University of Chicago Press.

Lang, Winfried. 1990/1991. "Is the Ozone Depletion Regime a Model for an Emerging Regime on Global Warming?" *UCLA Journal of Environmental Law and Policy* 9:161–174.

Langenau, Edward E., Jr., R. Ben Peyton, Julie M. Wickham, Edward W. Caveney, and David W. Johnston. 1984. "Attitudes toward Oil and Gas Development among Forest Recreationists." *Journal of Leisure Research* 16(2):161–177.

Lashof, Daniel A., and Dennis A. Tirpak, eds. 1991. *Policy Options for Stabilizing Global Climate.* Washington, D.C.: Climate Change Division, United States Environmental Protection Agency. (Page number citations refer to draft document, dated February 1989.)

Lave, Lester B. 1981. "A More Feasible Social Response." *Technology Review* (November/December):23–31.

Leopold, Aldo. 1970. *A Sand County Almanac, with Essays on Conservation from Round River.* New York: Oxford University Press, 1949. Reprint, New York: Ballantine Books.

Lessa, William, and Evon Vogt. 1972. *Reader in Comparative Religion.* 3d ed. New York: Harper and Row.

Lester, R. T., and J. P. Myers. 1989. "Global Warming, Climate Disruption, and Biological Diversity." Pp. 177–221 in *Audubon Wildlife Report 1989/90,* New York: Academic Press.

Lévi-Strauss, Claude. 1966. *The Savage Mind.* Chicago: University of Chicago Press.

Lewis, Paul. 1992. "U.S. Informally Offers to Cut Rise in Climate-Warming Gases." *New York Times,* 29 April A10.

Limbaugh, Rush. 1993. *The Way Things Ought To Be.* New York: Pocket Books.

Löfstedt, Ragnar E. 1992. "Lay Perspectives Concerning Global Climate Change in Sweden." *Energy and Environment* 3(2):161–175.

Löfstedt, Ragnar E. 1993. "Lay Perspectives Concerning Global Climate Change in Vienna, Austria." *Energy and Environment* 4(2):140–154.

Louis Harris and Associates, Inc. 1989. "Public and Leadership Attitudes to the Environment in Four Continents: A Report of a Survey in 16 Countries, Conducted for the United Nations Environment Programme." New York: Louis Harris and Associates, Inc.

Ludlum, David M. 1987. "The Climythology of America." *Weatherwise* 40:255–259.

Lyman, F., I. Mintzer, K. Courrier, and J. MacKenzie. 1990. *The Greenhouse Trap: What We're Doing to the Atmosphere and How We Can Slow Global Warming.* Boston: Beacon Press.

MacLean, Douglas. 1983. "A Moral Requirement for Energy Policies." In *Energy and the Future,* ed. Douglas MacLean and Peter G. Brown. Totowa, N.J.: Rowman and Littlefield.

MacLean, Douglas, and Peter G. Brown, eds. 1983. *Energy and the Future.* Totowa, N.J.: Rowman and Littlefield.

Mallinson, George G., J. B. Mallinson, L. Froschauer, J. A. Harris, M. C. Lewis, and C. Valentino. 1991. *Science Horizons.* Morristown, N.J.: Silver Burdett & Ginn.

Manabe, Syukuro, and Ronald J. Stouffer. 1993. "Century-scale Effects of Increased Atmospheric CO_2 on the Ocean-atmosphere System." *Nature* 364(15 July):215–218.

Manne, Alan S., and Richard G. Richels. 1989. "CO_2 Emission Limits: An Economic Cost Analysis for the USA." Paper presented at MIT Workshop on

Energy and Environmental Modeling and Policy Analysis, July. Also appears 1990 in *The Energy Journal* 11(2):51–74.

Manne, Alan S., and Richard G. Richels. 1993. "The EC Proposal for Combining Carbon and Energy Taxes: The Implications for Future CO_2 Emissions." *Energy Policy* 21(1):5.

Marshall, Lorna. 1957. "N!ow." *Africa* 27(3):232–240.

May, Robert M. 1976. "Simple Mathematical Models with Very Complicated Dynamics." *Nature* 261:459–467.

McCloskey, M., A. Caramazza, and B. Green. 1980. "Curvilinear Motion in the Absence of External Forces: Naive Beliefs about the Motion of Objects. *Science* 210:1139–1141.

McCloskey, M. 1983. "Intuitive Physics." *Scientific American* 284(4):122–130.

Merchant, Carolyn. 1992. *Radical Ecology*. New York: Routledge.

Meyer, Stephen M. 1993. "Dead Wood." *The New Republic* (2 August):12–15.

Milbrath, Lester W. 1984. *Environmentalists: Vanguard for a New Society*. Albany: State University of New York Press.

Milbrath, Lester W. 1989. *Envisioning a Sustainable Society: Learning Our Way Out*. Albany: State University of New York Press.

Mitchell, Robert Cameron. 1989. "From Conservation to Environmental Movement: The Development of the Modern Environmental Lobbies." Pp. 81–113 in *Government and Environmental Politics: Essays on Historical Developments Since World War Two,* ed. Michael J. Lacey. Washington, D.C.: Wilson Center Press.

Mitchell, Robert Cameron, Angela G. Mertig, and Riley E. Dunlap. 1991. "Twenty Years of Environmental Mobilization: Trends Among National Environmental Organizations." *Society and Natural Resources* 4:219–234.

Mohai, Paul. 1985. "Public Concern and Elite Involvement in Environmental-Conservation Issues." *Social Science Quarterly* 66(4):820–838.

Mohai, Paul. 1990. "Black Environmentalism." *Social Science Quarterly* 71(4):745–765.

Mohai, Paul, and Ben W. Twight. 1987. "Age and Environmentalism: An Elaboration of the Buttel Model Using National Survey Evidence." *Social Science Quarterly* 68(4):798–815.

Morgan, M. Granger, Baruch Fischhoff, Ann Bostrom, Lester Lave, and Cynthia J. Atman. 1992. "Communicating Risk to the Public: First, Learn What People Know and Believe." *Environmental Science and Technology* 26(11):2048–2056.

Morgan, M. Granger and Hadi Dowlatabadi, with Lester Lave, Mitchell Small, Edward Rubin, Baruch Fischhoff, and Paul Fischbeck. 1992b. "An Integrated Asssessment of Climate Change." Presentation at the annual meeting of the Society for Risk Analysis, 6–9 December, San Diego, Cal.

MORI (Market and Opinion Research International, Ltd.). 1990. "Home Insulation and the Environment." Research study conducted for EURISOL (Redbourn, Herts, U.K.), August.

Morrison, Denton E., and Riley E. Dunlap. 1986. "Environmentalism and Elitism: A Conceptual and Empirical Analysis." *Environmental Management* 10(5):581–589.

Muir, John. 1911. *My First Summer in the Sierra.* Boston: Houghton Mifflin. (Reprinted 1944.)

Nader, Laura, and Stephen Beckerman. 1978. "Energy as It Relates to the Quality and Style of Life." *Annual Review of Energy* 3:1–28.

Naess, Arne. 1989. *Ecology, Community and Lifestyle: Outline of an Ecosophy.* Translated and reviewed by David Rothenberg. Cambridge: Cambridge University Press.

Nash, Roderick Frazier. 1989. *The Rights of Nature: A History of Environmental Ethics.* Madison: University of Wisconsin Press.

Nassar, Sylvia. 1992. "Can Capitalism Save the Ozone?" *New York Times* 7 February, D2.

National Academy of Sciences. 1991a. *Policy Implications of Global Warming: Report of the Mitigation Panel.* Washington, D.C.: National Academy Press.

National Academy of Sciences. 1991b. *Policy Implications of Global Warming: Report of the Adaptation Panel.* Washington, D.C.: National Academy Press.

National Academy of Sciences. 1991c. *Policy Implications of Greenhouse Warming—Synthesis Panel.* Washington, D.C.: National Academy Press.

National Research Council. 1992. *Automotive Fuel Economy: How Far Should We Go?* Washington, D.C.: National Academy Press.

Newsweek. 1988. "Inside the Greenhouse," 11 (July): 17ff. (Special report consisting of three articles.)

Nietschmann, B., 1984, "Indigenous Island Peoples, Living Resources, and Protected Areas." Pp. 333–343 in *National Parks, Conservation, and Development: The Role of Protected Areas in Sustaining Society,* ed. J. A. McNeely and K. R. Miller. Washington, D.C.: Smithsonian Institution Press.

Nordhaus, William D. 1991. "To Slow or Not to Slow: The Economics of the Greenhouse Effect." *The Economic Journal* 101(6):920–937.

Norgaard, Richard B., and Richard B. Howarth. 1991. "Sustainability and Discounting the Future." Pp. 88–101 in *Ecological Economics: The Science and Management of Sustainability,* ed. R. Costanza. New York: Columbia University Press.

Oates, David. 1989. *Earth Rising: Ecological Belief in an Age of Science.* Corvallis, Ore.: Oregon State University Press.

Odum, Eugene P. 1969. "The Strategy of Ecosystem Development." *Science* 164(18 April):266ff.

Office of Technology Assessment (OTA). 1991. *Changing by Degrees: Steps to Reduce Greenhouse Gases*. OTA-O-482, February. Washington, D.C.: U.S. Government Printing Office.

Olsen, Marvin E., Dora G. Lodwick, and Riley E. Dunlap. 1992. *Viewing the World Ecologically*. Boulder, Col.: Westview Press.

O'Neill, R. V., D. L. DeAngelis, J. B. Waide, and T. F. H. Allen. 1986. *A Hierarchical Concept of Ecosystems*. Princeton, N.J.: Princeton University Press.

Paehlke, Robert C. 1989. *Environmentalism and the Future of Progressive Politics*. New Haven: Yale University Press.

Paine, R. T. 1966. "Food Web Complexity and Species Diversity." *The American Naturalist* 100:65–75.

Passel, Peter. 1989. "Cure for Greenhouse Effect: The Costs Will Be Staggering." *New York Times*, 19 November.

Passmore, John. 1974. *"Man's Responsibility for Nature: Ecological Problems and Western Traditions."* New York: Charles Scribner's Sons.

Peña, Devon. 1992. "The 'Brown' and the 'Green': Chicanos and Environmental Politics in the Upper Rio Grande." *Capitalism, Nature, Socialism* 3(1):79–103.

Peters, Robert L., and Thomas E. Lovejoy. 1990. "Terrestrial Fauna." Pp. 353–369 in *The Earth as Transformed by Human Action: Global and Regional Changes in the Biosphere over the Past 300 Years*, ed. B. L. Turner et al.

Peters, Robert L., and Thomas E. Lovejoy, eds. 1992. *Global Warming and Biological Diversity*. New Haven: Yale University Press.

Pickett, Steward T., and P. S. White, eds. 1985. *The Ecology of Natural Disturbance and Patch Dynamics*. Orlando, Fla.: Academic Press.

Plotkin, S. E. 1993. "Report on Reports: Automotive Fuel Economy: How Far Should We Go?" *Environment* 35(3):25–29.

Rappaport, Roy A. 1968. *Pigs for the Ancestors: Ritual in the Ecology of a New Guinea People*. New Haven: Yale University Press.

Raven, Peter H., and Edward O. Wilson. 1992. "A Fifty-Year Plan for Biodiversity Surveys." *Science* 258 (13 November):1099–1100.

Read, Daniel, Ann Bostrom, M. Granger Morgan, Baruch Fischhoff, and Tom Smuts. 1994. "What Do People Know about Global Climate Change? Part 2: Survey Studies of Educated Laypeople." To appear in *Risk Analysis* 4 (6).

Redclift, Michael R. 1987. *Sustainable Development: Exploring the Contradictions*. New York: Methuen.

Roberts, Leslie. 1991. "Report Nixes 'Geritol' Fix for Global Warming." *Science* 253(27 September):1490–1491.

Rolston, Holmes. 1988. *Environmental Ethics: Duties to and Values in the Natural World*. Philadelphia: Temple University Press.

Romney, A. Kimball, William H. Batchelder, and Susan C. Weller. 1987. "Recent Applications of Cultural Consensus Theory." In *Intracultural Variation*, ed. James S. Boster, special issue of *American Behavioral Scientist* 31(2):163–177.

Romney, A. Kimball, Susan C. Weller, and William H. Batchelder. 1986. "Culture as Consensus: A Theory of Culture and Informant Accuracy." *American Anthropologist* 88:313–338.

Ross, Michael. 1993. "Reconstructing Nature: A Brief History of Environmental Anxiety." *Praxis: Graduate Criticism and Theory,* 1–6. New Brunswick, N.J.: Rutgers University.

RSM (Research/Strategy/Management, Inc.). 1989. "Global Warming and Energy Priorities: A National Perspective." Survey commissioned by the Union of Concerned Scientists. Unpublished manuscript, November, available from UCS, Cambridge, Mass.

Sagoff, Mark. 1991. "Zuckerman's Dilemma: A Plea for Environmental Ethics." *Hastings Center Report* 21(5):32–40.

Sahlins, Marshall. 1972. *Stone Age Economics.* Hawthorne, N.Y.: Aldine de Gruyter.

Schelling, Thomas C. 1990a. "Global Environmental Forces." Pp. 75–84 in *Energy: Production, Consumption and Consequences,* ed. John L. Helm. Washington, D.C.: National Academy Press.

Schelling, Thomas C. 1990b. Public Remarks at Conference on Energy and the Environment in the 21st Century, Massachusetts Institute of Technology, 26–28 March.

Schnaiberg, Allan. 1994. *Environment and Society: The Enduring Conflict.* New York: Saint Martin's Press.

Schneider, Keith. 1992. "When the Bad Guy Is Seen as the One in the Green Hat." *New York Times* 16 February.

Schneider, Stephen H. 1988. "Doing Something about the Weather." *World Monitor, The Christian Science Monitor Monthly* 1 (3):28–37.

Schneider, Stephen H. 1989. "The Greenhouse Effect: Science and Policy." *Science* 243(10 February):771–781.

Schneider, Stephen H. 1990. Lecture at Princeton University, 1 March.

Shabecoff, Philip. 1987. "Dozens of Nations Reach Agreement to Protect Ozone." *New York Times,* 17 September, A1.

Shugart, H. H., and D. C. West. 1977. "Development of an Appalacian Deciduous Forest Succession Model and Its Application to Assessment of the Impact of the Chestnut Blight." *Journal of Environmental Management* 5:161–179.

Sills, David L. 1975. "The Environmental Movement and Its Critics." *Human Ecology* 3(1):1–41.

Simberloff, D. 1986. "Are We on the Verge of a Mass Extinction in Tropical Rainforests?" Pp. 165–180 in *Dynamics of Extinction,* ed. D. K. Elliott. New York: Wiley.

Simons, M. 1993. "Twelve Countries, 340 Million Shoppers, One Planet." *New York Times,* 11 April, Sec. 4, 5.

Smallwood, William L., and Edna R. Green. 1977. *Biology.* Morristown, N.J.: Silver Burdett & Ginn.

Smith, R. C., B. B. Prézelin, K. S. Baker, R. R. Bidigare, N. P. Boucher, T. Coley, D. Karentz, S. MacIntyre, H. A. Matlick, D. Menzies, M. Ondrusek, Z. Wan, and K. J. Waters. 1992. "Ozone Depletion: Ultraviolet Radiation and Phytoplankton Biology in Antarctic Waters." *Science* 255 (21 February):952–959.

Spencer, Roy W., and John R. Christy. 1990. "Precise Monitoring of Global Temperature Trends from Satellites." *Science* 247(30 March):1558–1562.

Spradley, James P. 1979. *The Ethnographic Interview.* New York: Holt, Rinehart and Winston.

Stern, Paul C. 1992. "What Psychology Knows About Energy Conservation." *American Psychologist* 47:1224–1232.

Stern, Paul C., and Thomas Dietz. 1994. "The Value Basis of Environmental Concern." To appear in *Journal of Social Issues.*

Stern, P. C., Thomas Dietz, and Linda Kalof. 1993. "Value Orientations, Gender, and Environmental Concern." *Environment and Behavior* 25(3):322–348.

Stern, P. C., and G. T. Gardner. 1981. "Psychological Research and Energy Policy." *American Psychologist* 36(4):329–342.

Stevens, William, K. 1991. "Summertime Harm to Ozone Detected over Broader Area." *New York Times,* 23 October A1, A11.

Stewart, Thomas R., Jeryl L. Mumpower, and Patricia Reagan-Cirincione. 1992. "Scientists' Agreement and Disagreement about Global Climate Change: Evidence from Surveys" April, 26 pp. Albany, N.Y.: Center for Policy Research, State University of New York.

Sundquist, Eric T. 1990. "Long-Term Aspects of Future Atmosphereic CO_2 and Sea-Level Changes." Pp. 193–207 in *Sea-Level Change,* Roger R. Revelle et al. Washington, D.C.: National Research Council and National Academy Press.

Taubes, Gary. 1993. "The Ozone Backlash." *Science* 260(11 June):1580–1583.

Taylor, Bron. 1991. "The Religion and Politics of Earth First!" *The Ecologist* 21(6):258–256.

Taylor, Bron. 1993. "Earth First!'s Religious Radicalism." Pp. 185–209 in *Ecological Prospects: Scientific, Religious, and Aesthetic Perspectives,* ed. Christopher Chapple. Albany, N.Y.: State University of New York Press.

Tribe, Lawrence. 1974. "Ways Not to Think about Plastic Trees: New Foundations for Environmental Law." *Yale Law Journal* 83(7):1315–1346. (Reprinted in *People, Penguins, and Plastic Trees: Basic Issues in Environmental Ethics,* ed. Donald VanDeVeer and Christine Pierce. Belmont, Calif.: Wadsworth Publishing Co., 1986.)

Tucker, William. 1982. *Progress and Privilege: America in the Age of Environmentalism.* Garden City, N.Y.: Anchor Press/Doubleday.

Turner, B. L., William C. Clark, Robert W. Kates, John F. Richards, Jessica T. Mathews, and William B. Meyer, eds. 1990. *The Earth as Transformed by*

Human Action: Global and Regional Changes in the Biosphere over the Past 300 Years. Cambridge, U.K.: Cambridge University Press.

Tversky, Amos, and Daniel Kahneman. 1982. "Judgment under Uncertainty: Heuristics and biases." Pp. 3–20 in *Judgment under Uncertainity: Heuristics and Biases,* ed. D. Kahneman, P. Slovic, and A. Tversky. Cambridge, U.K.: Cambridge University Press.

U.S. News & World Report. 1963. "Is Man Upsetting the Weather?" 11 November, 46–49.

Vitousek, Peter M., and Lawrence R. Walker. 1989. "Biological Invasion by *Myrica Faya* in Hawai'i: Plant Demography, Nitrogen Fixation, Ecosystem Effects." *Ecological Monographs* 59(3):247–265.

Walker, James C. G., and James F. Kasting. 1992. "Effects of fuel and forest conservation on future levels of atmospheric carbon dioxide." *Palaeogeography, Palaeoclimatology, Palaeoecology (Global and Planetary Change Section)* 997:151–189.

Weller, Susan C. 1987. "Shared Knowledge, Intracultural Variation, and Knowledge Aggregation." *American Behavioral Scientist* 31(2):178–193.

Weller, Susan C., A. Kimball Romney, and Donald Orr. 1986. The Myth of a Sub-Culture of Corporal Punishment. *Human Organization* 46:39–47.

West, Patrick C., and Steven R. Brechin, eds. 1991. *Resident Peoples and National Parks: Social Dilemmas and Strategies in International Conservation.* Tucson: The University of Arizona Press.

White Jr., Lynn. 1967. "The Historical Roots of Our Ecological Crisis." *Science* 155(10 March):1203–1207.

Williamson, Hugh. 1771. "An Attempt to Account for the Change Observed in the Middle Colonies in North America." *Transactions of the American Philosophical Society,* 1st ed. (Cited in Ludlum 1987.)

Wilson, Edward O. 1989. "Threats to Biodiversity." *Scientific American* 261(3):108–116.

Wilson, Edward O. 1992. *The Diversity of Life.* Cambridge, Mass.: Harvard University Press.

Wilson, Edward O., and Stephen R. Kellert. 1993. *The Biophilia Hypothesis.* Washington, D.C.: Island Press/Shearwater Books.

World Commission on Environment and Development. 1987. *Our Common Future.* New York: Oxford University Press.

World Resources Institute. 1992. *World Resources 1992–93: A Guide to the Global Environment.* New York: Oxford University Press.

Worster, Donald. 1977. *Nature's Economy: A History of Ecological Ideas.* New York: Cambridge University Press.

Worster, Donald. 1990. "The Ecology of Order and Chaos." *Environmental History Review* (Spring/Summer):1–18.

Wynne, Brian. 1991. "Knowledges in Context." *Science, Technology and Human Values* 16:111–121.

Wynne, Brian. 1992. "Misunderstood Misunderstanding: Social Identities and Public Uptake of Science." *Public Understanding of Science* 1:281–304.

Index